OPEC OIL

OPEC Oil

Loring Allen

University of Missouri—St. Louis

 Oelgeschlager, Gunn & Hain, Publishers, Inc.
Cambridge, Massachusetts

110850

International Standard Book Number: 0-89946-002-x

Library of Congress Catalog Card Number: 79-19284

Printed in the United States of America

Library of Congress Cataloging in Publication Data

Allen, Loring.
    OPEC oil.

    Bibliography: p. 247
    Includes index.
    1. Organization of Petroleum Exporting Countries. 2. Petroleum industry and trade. I. Title.
HD9560.1.O66A69   382'.42'282   79-19284
ISBN 0-89946-002-x

To Jeanne

Contents

Preface

Few words in the language hold more portent than "OPEC," a new word that stands for the Organization of Petroleum Exporting Countries. OPEC consists of thirteen countries that own 68 percent of existing world oil and supply 84 percent of annual crude-oil exports. In 1973 OPEC took control of the crude-oil industry away from the oil companies and has since regulated production and set the price. Its actions in this decade have increased the price of crude oil more than eleven-fold. It has upset the world power balance, worsened inflation and recession, and forced oil-consuming countries to make unwanted changes, while greatly enhancing its members' welfare and prospects of development.

OPEC, along with the oil companies and government energy policies, has earned the emnity of the American people. Long accustomed to low oil prices and secure supplies, consumers now confront high and rising prices and uncertain supplies. They reason that government bungling, as well as oil-company and OPEC greed, must be at fault. The oil companies respond that they must buy OPEC oil at its price or have no oil, since they can produce no more at home. Yet they profit from OPEC oil. Government officials call OPEC an unwholesome cartel and bemoan the price increases.

Yet the government treats OPEC members, as well as the oil companies, with kid gloves and shuns stern conservation measures. So loud are the complaints against OPEC that seldom does the case for OPEC reach American shores.

This book seeks to add perspective to the debate over OPEC by tracing its development and indicating its role in world oil. I have tried to focus on OPEC, seeing the United States and oil-importing countries only with peripheral vision. In this way, OPEC emerges as a natural response to world economic growth, the conditions of world oil, and the struggle for power over its supply. Americans have no reason to feel grateful for OPEC's price behavior. But Americans should understand how and why these thirteen small countries behave the way they do and what their actions imply for oil, the standard of living, and economic growth.

Some regard OPEC actions as morally reprehensible. It is wrong, they say, for a monopoly to exploit consumers. If those who stand on high principle judged the exercise of power by others with the standards of their own behavior, they might be slower to find fault. When Americans and their oil companies controlled world oil, the moralists were silent, even though cheap oil degraded peoples in oil-producing countries and blocked their progress. Perhaps it is best to recognize that world economic relations are neither right nor wrong. They just are. And they serve interests. The world oil industry has long served American interests and still does, despite the changes wrought by OPEC. But now, world oil serves OPEC's interests better than before.

My purpose is not to criticize or judge oil consumers, their governments, the oil companies, or OPEC. All have behaved well and badly at different times. All have acted according to their own interests. Americans should no more expect Saudi Arabia to deny its interests in their favor any more than Saudi Arabians can expect Americans to pay a trillion dollars in taxes in the next forty years to develop the Saudi Arabian economy. Oil companies are now no more willing to forego growth and profits to benefit consumers than they were willing a decade ago to benefit oil-exporting countries out of their own pockets. The blame for the world's oil problems lies not in evil men behaving with malevolent intent, but rather in good men pursuing their legitimate but conflicting interests.

Let me not conceal my own wholly nonscientific sentiments. I admire those five spunky little countries—Iran, Iraq, Kuwait, Saudi Arabia, and Venezuela—who put their heads together in 1960 to found OPEC. They had felt the lash of the oil companies and decided to do something about it. They assaulted the world's greatest assemblage of private economic power that was serving the interest of the richest peoples on earth. That took guts.

The members of OPEC patiently learned the oil business and applied the little power they had to wrest small benefits. When the conditions of world oil changed in 1970, they were ready. They did not hesitate to use the enhanced power that galloping consumption and limited supplies conferred on them. In 1973 they had the audacity and the ability to take control of the industry, thereafter exploiting consumers and the companies as consumers and the companies had long exploited them. It is poetic justice.

OPEC members are managing the crude-oil industry with greater prudence than that exercised in the past by the companies and consumers. That their management offers at least the opportunities for them to develop economically is welcome. That their management imposes long deferred short-term costs and painful adjustments on oil-importing countries is unfortunate. That their management conserves a vital nonrenewable resource and provides time and incentive for the world to change to substitutes can be a great windfall gain for everyone. But make no mistake—OPEC runs the oil industry to serve its own interests.

The secret of OPEC's success is cooperation. The self-serving conflict of markets often produces, in my view, inequitable results. Cooperation, by reconciling conflicting interests, can sometimes help to erase inequities. I believe it has in the OPEC case. Of course, I am no happier about paying more for gasoline than any other American, nor am I pleased at the unpleasant changes that all Americans must make. But perhaps now the United States, taking a leaf from OPEC's book, can begin cooperating with other oil-importing countries and with OPEC to their mutual benefit.

In this book I have tried to tell the story of why and how OPEC developed, its achievements, its impact, and its prospects. OPEC was born and lives in controversy. No telling of its story can avoid controversy. Well represented in the bibliography are those who support consumer interests. Others espouse the OPEC cause. This book doesn't take sides in this issue. Rather, it purports only to examine the origin, performance, and consequences of OPEC.

Mr. Hamid Zaheri and Mr. John Hunter of OPEC and Dr. Ibrahim Shihata of the Opec Special Fund, as well as others in OPEC, made my visit to their Vienna headquarters in 1978 useful and memorable. I am especially grateful to H.E. Ali M. Jaidah, OPEC Secretary General, for an interview that appeared in *Worldview* in March 1979. My colleagues and students at the University of Missouri-St. Louis and Universidad Simón Bolívar in Caracas, Venezuela, have contributed more than they know. The University Center for International Studies and Office of Research provided financial assistance. No one is to blame but myself, however, for the remaining errors or benighted interpretations.

The first chapter relates the circumstances and events that transferred the control of world oil from the oil companies to OPEC and the immediate reaction. The oil revolution, however, is understandable only in the context of the development of the industry. This sketch is followed by chapters that deal with OPEC in the 1960s and with the three years of the 1970s leading up to the denouement. Chapters follow that detail OPEC actions for the rest of the 1970s, how OPEC has organized the oil market, and OPEC's earnings. Two chapters show OPEC's special relations with the less-developed countries. The book ends with a glimpse into the future, a look at the stability of OPEC, and an estimate of OPEC's place in the world.

St. Louis, Missouri
July 1979

Chapter 1

The OPEC Revolution

When Saudi Arabia and the world's leading oil-exporting countries confronted Exxon and the world's largest oil companies to negotiate the price of oil in October 1973, the companies blinked first. The events of the next two and one-half months changed forever the world crude-oil industry and heralded the beginning of the end of the oil age. The price of crude oil quadrupled. The control of price and production shifted from the oil companies and oil-importing countries to the owners and exporters of oil, the members of the Organization of Petroleum Exporting Countries (OPEC). When the dust settled in early 1974, the "quarreling collection of camel sheikhdoms and banana republics," as the *Washington Post* later called them, had taken over world oil.

Four years of turmoil in the oil market preceded this last price negotiation. OPEC members jockeyed for position with the companies for more than a decade, losing some skirmishes and winning others. Beginning in 1970, events move swiftly in OPEC's favor. The oil glut of the 1960s disappeared, replaced by growing shortages, as the United States and other industrial countries demanded more and more crude oil. In 1970 Libya coerced its oil companies to raise its crude price and to pay more taxes. Not to be outdone, Iran in 1971 obtained for the Persian Gulf countries an even higher price.

1

Foreign offices in London, Paris, Tokyo, Bonn, and Washington wrung their hands but had no alternative but to pay.

Six oil ministers, all from the Persian Gulf countries, represented the eleven members of OPEC in the October meeting. The other members had agreed to abide by the price that the six negotiated. Saudi Arabia and Iran, newly installed as the world's largest oil exporters, led the OPEC delegation that also included Kuwait, Iraq, Abu Dhabi, and Qatar. Note the one non-Arab country—Iran—amid the five Arab countries. What followed was not exclusively an Arab show.

Five company negotiators, representing the "Seven Sisters"— Exxon, Shell, British Petroleum, Texaco, Mobil, Gulf, and Socal—and seventeen other large companies, sat on the other side of the table. By choice of the companies and governments alike, no representatives of the governments of oil-importing countries accompanied the senior oil executives. The meeting place was Doktor Karl Lueger Ring 10, the austere old headquarters of OPEC across the street from the University in the Inner City of Vienna. Both OPEC and the companies knew that the meeting was important. Both sides had girded for a tough bargaining session.

At stake was the posted price of crude oil. For the "marker" grade—Arabian light—that price before the meeting stood at $3.01 a barrel. This price was only one of many prices for crude oil. It was the price that the companies announced, indicating their readiness to supply oil at that price. In the distant past it had been close to the actual sale price, but in the years leading up to 1973 the posted price had exceeded the market price by one-fourth to one-half. The market price responded to the ebb and flow of supply and demand. It also included discounts and premiums that varied with the specific gravity, location, and other physical and chemical properties of the oil, as well as the nature of the market through which the oil flowed.

The posted price was critical because the oil companies and the oil-exporting countries had agreed to use it to calculate the companies' taxes and royalties. The royalty is the share of oil—in most cases in the Middle East, 12.5 percent—that belonged to the producing nation. Anciently, it was the king's share of mining production. It was not a tax, although commonly so treated. The taxes imposed by the producing countries, primarily income taxes, were based on the companies' net revenues. Rather than multiplying the oil produced by market prices that varied from week to week, the companies multiplied the more stable posted price by the amount of oil produced.

Both the market and posted prices had climbed in the four years preceding the Vienna meeting. In 1970, before Libya pressured the companies to raise it, the posted price was $1.80 a barrel. In February 1971, as a result of the agreement between the companies and OPEC members at Teheran, it went up to $2.10. In 1971 it rose twice by small amounts, and in 1972 it continued to creep up. Each small jump reflected a new agreement between the companies and OPEC. Between early 1970 and mid-1973, the posted price rose 67 percent.

Before October 1973 the oil companies had always set the posted price. The companies' purpose in announcing the price and changing it from time to time was to indicate the state of the market and to signal supply and demand changes. The posted price and the market price were interdependent, each helping to determine the other. Changes in the posted price influenced costs and the companies' willingness to supply oil. Changes in the market price influenced the companies' profit positions. The oil companies tried to keep both prices stable by controlling supplies and influencing market demand.

If inventories accumulated and demand slackened, resulting in the weakness of the market price, some oil companies—the large ones—first restricted production to try to strengthen the price. But competition—increasing production by other enterprises—sometimes defeated these efforts to shore up a declining price. On the other hand, a booming market tended to induce all the companies to open the supply tap more. In either case, the market price moved. However, only major shifts in the market price, or pressure from OPEC, forced the companies to raise the posted price, since they would then have to make higher tax and royalty payments.

Members of OPEC had always pressed the companies to increase the posted price. The increases of 1970-1973 responded to OPEC's urgings, helped along by the rising market price that in turn resulted from growing demand and tighter supplies. The movement of the market price, when crude was in short supply in the early 1970s, augmented the OPEC members' resolve to keep the posted price moving up as well. These market-price increases also weakened the oil companies' ability to argue that the posted price should be held down.

As the market price advanced steadily in the 1970s, the gap between the market price and the more sluggish posted price narrowed. For the first time, in mid-1973, the market price exceeded the posted price. The oil companies were thus paying royalities and taxes calculated with a price lower than what they were actually receiving. OPEC members regarded this as an intolerable padding of company profits, which they argued had tripled in the previous three years. The OPEC share of the oil dollar was deteriorating, even though the dollars earned were increasing because of the larger volume of oil, as well as the modest posted-price increases. OPEC members intended to reverse this anomaly.

The Teheran Agreement of 1971 still governed the posted price for the Persian Gulf oil states. Some other members nominally set their own prices, but they could not move much out of line with the Gulf price. The agreement, in establishing the posted price, had also contained an escalator clause that allowed for modest posted-price increases because of inflation in the industrial countries.

The Teheran Agreement had not contemplated, however, the declining value of the dollar, the currency that the countries and the companies use in oil transactions. When the dollar toppled in December 1971, OPEC persuaded the companies to boost the posted price 9 percent to offset the reduced value of the dollar. The first Geneva agreement, in January 1972, also included a formula linking the posted price to a group of currencies to assure the stability of the purchasing power of OPEC earnings. In June 1973, following the second cut in the dollar's value, another Geneva agreement raised the posted price again and revised the currency formula.

The Teheran Agreement had also not foreseen the rapidly advancing oil market in which the market price overran the posted price. Although the agreement was designed to last for five years, by mid-1973 both OPEC and the companies recognized that events had overtaken it. The 35th OPEC Conference in mid-September summoned the companies to the meeting in Vienna to abrogate the agreement by setting a price outside its framework.

THE VIENNA STAND-OFF

OPEC members came to the meeting determined to increase the posted price substantially. Pre-meeting statements by the oil ministers, dismissed by the press and the companies as bluster, were unmistakably angry. The oil companies accepted the need to raise the posted price to meet the complaints of the exporting countries. A market price higher than the posted price, the companies recognized, prejudiced OPEC's interests. But the companies wanted a small increase. A large increase, effected by raising tax and royalty costs and possibly reducing product sales as these costs were passed through, would unsettle the market and might imperil company profits.

The only question on the table at the October meeting was the size of the price hike. Before the meeting, both sides anticipated difficult but amicable negotiations. The negotiators at the table were all old friends and antagonists, veterans of other stormy negotiations. Following the hard bargaining, both the oil ministers and the company executives expected that the companies, as in the past, would announce the new posted price that all had agreed on. The market would fret for a while and then calm down.

Later congressional investigations revealed much of what happened at the Vienna meeting. The most detailed study was that of the Church Committee reported in the many volumes of the *Hearings before the Subcommittee on Multinational Corporations on the Multinational Petroleum Companies and Foreign Policy* (Government Printing Office, 1974). Anthony Sampson, an English journalist, interviewed many of the principals and reported the meeting in his *Seven Sisters* (Viking Press, 1975).

The chief negotiator for the oil companies was George T. Piercy, senior executive vice-president for Exxon, the world's largest oil company. He had checked with the London Policy Group, the organization that all the oil companies had formed to deal with OPEC members on this occasion. He also had a waiver from the U.S. Department of Justice, permitting joint negotiations without fear of subsequent antitrust legal action. Piercy spoke for one of the largest conglomerations of private economic power ever assembled.

Piercy wanted an increase of no more than 15 or 20 percent. He could accept no more than 25 percent, the increase established by the London Policy Group as the upper limit. To go beyond that figure, Piercy had to consult with the companies. The companies then possibly would have to consult with the governments of the oil-importing countries. The oil executives thought that they could cool OPEC's ardor and keep the price within the limit.

Sheikh Yamani was the principal spokesman for the Persian Gulf oil ministers. His country exported more oil and had more reserves than any country in the world. He had no specific instructions, but he and his king were conservatives on the price issue. His colleagues, including Dr. Jamshid Amouzegar, Iran's oil minister, were more ambitious. They wanted at least a doubling of- the posted price. Revenue needs of members, the market price in excess of the posted price, and the belief that company profits were excessive undergirded the OPEC stance. Although the ministers may not have been as well informed about market conditions as the company executives, their urgency for a large increase was strong.

Some of the radical Arab countries, including Iraq and Libya, added pressure for strong price action. They not only wanted a large increase but also argued that OPEC members—all of them in concert—should dictate their own prices, independent of what the companies wanted. At the September OPEC Conference many members had favored a Venezuelan resolution—ultimately defeated—to set the posted price without even consulting the companies.

The negotiators were old adversaries. Piercy, a chemical engineer, up from the ranks of Exxon, was the company's Middle Eastern representative and a director of the company. Although he had been only a few years on the job in the Middle East, he knew the oil business. He recognized that in the prosperous oil market, the bargaining power of the companies was slipping and that of OPEC members was climbing. But he did not question that the oil companies were still king of the mountain.

Sheikh Yamani, a lawyer, had been the oil minister of Saudi Arabia since 1962. He succeeded the volatile Sheikh Abdullah Tariki, who had been instrumental in the founding of OPEC. Sheikh Yamani had the confidence of King Faisal, and like his sovereign, was a friend of the United States. Born in Mecca, he earned the title sheikh after graduating from New York University and working at the Center for International Legal Studies at Harvard University. The young sheikh—only forty-three in 1973—had become Mr. OPEC. He knew the companies and the oil business, and for him business came before friendship. More than anyone else, he had been responsible for the OPEC oil policy, enunciated in 1968, that emphasized price and production control. Now was his chance to exploit OPEC's bargaining advantage.

Fate intervened on the side of OPEC just two days before the meeting began. On Saturday, October 6, 1973, the Egyptian Second and Fourth Armies jumped off across the Suez Canal and fanned out into the Sinai Desert. At the same moment 700 Syrian tanks broke through the barbed-wire cease-fire line and lumbered across the Golan Heights toward Kuneistra and Rafid in Israel. The Yom Kippur War, the fourth major outbreak of armed conflict between the Arabs and Israelis, erupted. In Vienna the Arabs in the OPEC delegation buzzed with excitement and war talk.

The tickers brought news of initial Arab victories and of captured and destroyed Israeli military equipment. Egypt and Syria hoped for a quick recapture of some of their lost lands and for an international conflict that would lead to a settlement with Israel in their favor. In Vienna, oil and war mixed in the Arabs' discussions. President Anwar Sadat of Egypt had said in April 1973 that oil would be a weapon in any future Arab-Israeli conflict. At the secret meeting in August 1973, King Faisal of Saudi Arabia had assured President

Sadat that he would restrict oil exports if American policy toward the Arab cause did not improve.

The oil companies had known that the war was coming but not when. Tension had been rising in the Middle East for more than a year. Arab countries had spent vast sums on arms and military training. The Arabs were more united than ever in their determination to regain the territory lost in the humiliating Six-Day War of 1967. Saudi Arabia had finally lined up with those who accepted war as the only solution. The companies and their home governments knew that war was in the making.

As early as May 1973, King Faisal had told Frank Jungers, president of the Arabian American Oil Company (Aramco), that Saudi Arabia would support Egypt if war were to break out. Four of the Seven Sisters—Exxon, Texaco, Gulf, and Socal—owned Aramco. These and other oil companies, sensing the coming trouble, had stepped up production all through 1973. The companies knew that they might face production cutbacks and in any case that renewed warfare would damage their interests. Home governments feared the coming conflict but felt powerless to resolve it or to disconnect it from oil.

But the problem in Vienna was the price of crude oil. Neither side wanted to mix the war and the oil price decision. The Arabs restrained themselves because they were acting as members of OPEC. And OPEC was an economic organization representing not only seven Arab states—none of which were belligerents in the war—but also four non-Arab states. Nevertheless, the war hovered in the background, united the Arab delegates, and made the companies uneasy.

On the first day of the meeting each side laid out its position. The oil company executives offered 15 percent, less than $3.50 a barrel. The OPEC oil ministers countered with 100 percent, more than $6.00 a barrel. The huge gap stunned the executives, who thought the OPEC demand outrageous. The ministers thought the companies were pinching pennies. Sheikh Yamani's demand, several times what the companies had anticipated, would upset the market, they feared, and would be wholly unacceptable to the governments of the oil-importing countries.

Crude at $6.00 a barrel was inconceivable to the companies. They set out their views of market conditions in painstaking detail, hoping to deflate the aspirations of the oil ministers. Penny by penny they analyzed the impact of price increases on the market, proving irrefutably to themselves, but not to the ministers, that their demand was impossible. Although impressed by the companies' arguments, the oil ministers were not as concerned as the companies were with how the oil market might react.

OPEC had made its own market analysis and the ministers believed that the $6.00 price was feasible. They realized that it would disturb the market and might reduce the amount that consumers bought, but they did not believe that the companies would dare pass along the full price increase. Although the increase would be painful, OPEC thought the companies would permit profits to fall and product sales to dip only slightly. OPEC was confident that the small production losses would be more than offset by the price increase, so that their revenues would go up about the same as the price.

The OPEC ministers thought along two lines: the market and equity. They believed that $6.00 a barrel was a fair price. To the companies, oil was a business and their only interest was to make money from its sale in a stable, growing, and predictable market. To OPEC members, oil was not only a business, in which they shared the aims of the companies, but it was also their principal or only resource for developing their economies and satisfying the aspirations of their peoples. The oil ministers insisted upon a fair return for the alienation of their countries' wasting asset—oil—regardless of the market.

The second gathering of the negotiators brought only modest changes in attitudes. The oil companies went up a bit and the oil ministers gave some ground. The two sides remained a world apart, $3.75 to $5.00 a barrel. The company representatives and OPEC ministers wrangled all day over the effect on the market as well as the impact on demand, government revenues, company profits, inflation, and growth. By the end of the day Piercy made it clear that the companies could not go beyond the 25-percent limit. Sheikh Yamani made it equally clear that the companies' offer was inadequate and that $5.00 oil was the minimum that OPEC would accept.

The delegates of the companies, now desperate but still hoping for an agreement, called their home offices and the London Policy Group. They responded as the companies always had: when confronted with defeat or an awkward situation, delay. They told Piercy that to go beyond the 25-percent limit, they would have to consult with their government. That took time. Piercy concluded that another meeting would be futile. Yamani and Piercy met privately late at night at the Vienna Intercontinental. Piercy relayed the answer of his companies and asked for a recess of two weeks to get new instructions.

Sheikh Yamani, on behalf of the oil ministers, refused to budge. To him, two weeks for new instructions was just another delaying tactic of the companies. He was noncommital on the renewal of the negotiations later, suggesting that the time for unilateral action by OPEC perhaps had come. He told Piercy that the company executives should stay tuned to their radios for news of the price of crude oil.

The companies had tied Piercy's hands. They didn't believe that OPEC would do what it was threatening. Once before, at Teheran, the companies had weakened in the face of OPEC threats. It had cost them dearly. This time they decided to call OPEC's bluff. The meeting was over. Yamani flew back to Riyadh that night. Never again would OPEC members negotiate the price of oil with the companies.

THE KUWAIT MEETINGS

In the first days of the war, Abdul Rahman Atiqi, Minister of Oil and Finance of Kuwait, had called a meeting for October 17 of the Organization of Arab Petroleum Exporting Countries (OAPEC). This group, founded in 1968, had no formal connection with OPEC, but all the Arab OPEC members were also members of OAPEC, as were other Arab countries, including Syria and Egypt, both belligerents in the war. The purpose of the meeting was to discuss the use of Arab oil as a weapon in the war against Israel. When the meeting of OPEC ministers with the companies in Vienna broke down, the ministers decided to meet among themselves in Kuwait on October 16, the day before the OAPEC meeting.

The OPEC ministers then decided to set the posted price of marker crude oil at $5.12 a barrel. This 70 percent increase was the minimum price that they had insisted upon at the Vienna meetings. It was lower than some OPEC ministers, who at the earlier meeting with the companies wanted a doubling of the price. It reflected the conservatism of Sheikh Yamani and Saudi Arabia, as well as the views of the companies on the state of the oil market.

The new price was not wholly arbitrary. Before 1971 the posted price had been about 40 percent higher than the market price for some time. In mid-1973 the market price was about $3.65 a barrel, about 20 percent higher than the posted price. The October 16 posted price was just 40 percent higher than that market price, restoring the old price gap. The ministers, in setting the price, intended to reestablish and to maintain the traditional differential between the posted and market prices. The logic of OPEC's move was lost on those who had to pay for the oil at the new price.

The price increase was unilateral. The exporting countries of the Persian Gulf simply announced the new posted price and started to collect the companies' tax and royalty obligations using the new price mandated by law. Except for the few instances in which individual OPEC members such as Venezuela had set their own prices, this was the first time that members had established among themselves a new posted price, independent of the wishes of the companies. The companies had no alternative but to pass most of the new taxes and royalties along to consumers.

How could the OPEC ministers be sure that the posted price that they had established would bear the appropriate relation to the market price? They couldn't, in fact, be certain. Setting the new posted price implied the ability and willingness to control production, restricting it to the amount that would move through the market at a price consistent with the posted price. When the companies had fixed the price earlier, they had been able to determine production as well.

The events of the early 1970s had transferred the ability to control production to OPEC members. Saudi Arabia, Kuwait, and other Persian Gulf Arab countries had demanded and received part ownership in their operating companies. Iraq had nationalized its oil industry. The shah had made it clear that Iran's industry, fully nationalized in 1973, would now determine how much the private consortium could buy. Libya had nationalized much of its industry. Algeria was taking over French oil interests. These and other countries regulated the industry to the degree that the countries, not the companies, in fact, could determine production and exports. All members made it clear that the private oil companies must serve national interests. By October 1973, members controlled the amount produced.

Still, everything depended on the acceptance of the price by all OPEC members and their cooperation in allowing only sufficient production to justify the new price. The OPEC members had no collective production control method, having failed to install one several times in the 1960s. If too much oil came onto the market, the market price would differ too greatly from the posted price. The oil companies might be able to recapture price control.

The continuation of OPEC price control depended upon the voluntary restraint of the members. Above all, it meant that Saudi Arabia, which alone exported more than one-fourth of all OPEC oil and possessed more than one-third of its reserves, would set its production so that the market price would not drift downward from the mid-1973 level.

In setting the price, Saudi Arabia and the other Arab OPEC members knew, and Iran could easily guess, what was going to happen the next day. OAPEC planned to deploy its oil weapon, cutting Arab production and embargoing shipments of oil to countries supporting Israel. The supplies of crude oil moving into world markets would decline even with modest cutbacks and an imperfect embargo. The market price would remain stable or even more probably move up, supporting the new posted price. When the OPEC meeting in Kuwait was over on October 16, Iran's minister departed hastily so that the events of the days following would not carry the OPEC stamp.

The October 16 meeting on the posted price was the watershed for world oil. Oil companies and oil-importing countries recognized that the power to make decisions on price and production had passed into OPEC's hands. The OPEC members asserted their authority and the companies and consumers accepted it. Their authority rested on the inability and unwillingness of the companies and oil importers such as the United States to control demand or challenge the imperfect control—dependent on OPEC voluntary restraint —of supply by adding to OPEC production. OPEC control created a new crude-oil market in which OPEC members, not the companies, were the suppliers.

Only time would tell how the new market would work. At the moment, with government-ordered cutbacks and an embargo hard on the heels of the price takeover, the market would accept the orders of the new suppliers and respond with higher prices. Had the companies or the United States and other importers announced that the price takeover was illegitimate and that they would not accept it, the company-dominated world oil market might have reemerged after the embargo and cutbacks were over. To be credible, however, oil importers would have to find sufficient supplies to meet their demand at a lower price. They couldn't and didn't.

Strangely enough, the OPEC price initiative was lost in the war news, the production cutbacks, and the embargo. Subsequently, many myths have sprung up surrounding the new OPEC price. Many people believe that OPEC forced the companies to increase the price. Not so. The OPEC members, on their own, posted the new price. Others imagine that the oil companies raised the price in order to profiteer, using OPEC as the excuse. Not so. The companies had resisted the price increase, but had to accept it. Many suggested that the Arabs raised the price. Not so. OPEC, Arabs and non-Arabs alike, raised the price. Some believe the State Department at fault. No so. Supply and demand and those that control them, not diplomatic overtures, set the oil price.

THE EMBARGO

Israel's losses early in the hostilities jeopardized its capacity to continue the war. Urgently needing military resupply, Israel appealed to the United States. Saudi Arabia, the most reliable friend of the United States among Arab countries, sent its foreign minister, Omar Saqqaf, to plead directly with President Richard Nixon for United States' nonintervention. The mission failed. Not only the president, but also the Congress and most of American public opinion supported helping Israel. The United States began to resupply Israeli forces and put its Sixth Fleet on alert.

Feeling rebuffed by the United States, Saudi Arabia made good its pledge to its Arab neighbors. It not only supported but also spearheaded the use of oil as a weapon. On October 18, Saudi Arabia ordered a 10-percent cutback in production. Frank Jungers, head of Aramco, heard about it on the radio. Three days after the Kuwait meeting, Sheikh Yamani told Jungers to cut Aramco production 25 percent and to stop all shipments of oil to the United States and its facilities, including its armed forces, and to the Netherlands. Jungers said later that "the only alternative was to ship no oil at all" (*Multinational Hearings*, Part 7, p. 517ff).

Most other Arab countries quickly fell into line, some taking an even more hard-nosed attitude. By October 22 all Arab producers except Iraq were embargoing shipments to the United States and had ordered cutbacks. Some also embargoed the Netherlands and other countries regarded as too sympathetic to Israel. The effect was not immediate, however, since shipments at sea continued to their preembargo destinations. Still, the oil-importing countries began to worry about supplies for the winter months ahead.

As the American resupply of Israel continued, Arab countries tightened the embargo. More important, they took measures to make the embargo more effective, plugging leaks, establishing destination rules, policing the embargo, tracking shipments, and trying to prevent sharing. Sheikh Yamani's computers worked overtime, following the trail of tankers and watching deliveries, destinations, and inventories. At the meeting on November 5, the Arab countries decided on a uniform 25-percent cutback. By now it was clear that the 1973 embargo was not to be a repetition of the 1956 and 1967 fiascos. Arab unity was holding, and the companies and consuming countries seemed almost powerless.

The production cutbacks began to influence the policies of oil-importing countries. The essays by James McKie (United States), Romano Prodi and Alberto Clô (Europe), and Yashi Tsurumi (Japan) in *The Oil Crisis*, R. Vernon, editor (Norton, 1976), show the response. Although the American pro-Israel stand did not deteriorate, Secretary of State Kissinger did make frantic efforts to arrange a cease-fire, hoping to end the hostilities and the embargo. The United States reaffirmed its support of United Nations Resolution 242 that favored Arab territorial claims, but still continued the military resupply of Israel. The European Economic Community capitulated, issuing on November 6 a strong statement supporting Resolution 242 and the Arab stand on their lost lands. Japan and other European countries announced that they accepted the Arab view. Great Britain, France, and others began to negotiate separate deals with Arab countries, exchanging arms and capital goods for oil.

At the summit meeting at the end of November, Arab countries accepted the Saudi Arabian proposal to continue to reduce exports until the cut in government revenues of each Arab country equalled 25 percent of 1972 revenues. This implied a cut in production of at least 75 percent. It was mainly window dressing, but it showed that the Arabs were intent, at this stage, on perservering. Saudi Arabian production had declined 27 percent between September and November. Kuwait's production was down 30 percent. All were earning more dollars than ever, however, because of the higher market price.

Iran increased its production slightly during the embargo. Venezuelan production increased a bit during the fourth quarter and its shipments to the United States went up. Iraq did not officially join the embargo, but its production did not increase because of damaged shipping facilities. Overall, OPEC members' production declined by more than 12 percent between September and November. The production cutback for those Arab countries participating in the embargo was 26 percent.

Just as the embargo and cutbacks seemed to be accelerating and Arab unity was still holding firm, the effectiveness of the embargo came into question. Leaks, the special deals between Arab and European countries outside the embargo, and sharing seemed to be undermining it. The higher price was also cutting consumption and the oil-importing countries had rushed to legislate conservation measures. Voluntary demand restraint—carless Sundays, carpooling, bicycling, and even walking—reduced consumption.

Even more important, it was becoming clear that the heavy

preembargo shipments by the oil companies had prepared the oil-importing countries to withstand the embargo. Saudi Arabia, for example, had produced 36 percent more oil in the first nine months of 1973 than in the same period of 1972. With bulging crude inventories and strong conservation, Europe and America battened down the hatches, ready to weather the storm. Short of intervention in the Middle East or a reversal of policy that was impossible under the circumstances, the United States and other oil-importing countries had no alternative.

COMPANIES AND GOVERNMENTS

The oil companies attracted as little attention to themselves in this period as possible. As large multinational corporations, they interpreted their task as survival, no matter what happened. They felt that their assets—concessions, equipment, raw materials, and markets—were hostage to the Arab countries. They behaved to satisfy the Arab demands on them. They obeyed the letter of the embargo and production cutback decrees, regardless of the impact on their home countries.

The oil companies did try to circumvent the embargo as much as they could. They shipped oil to make it available to the United States and the other importing countries against the spirit of the Arab embargo. With the experience of the 1956 and 1967 embargoes, they knew many tricks to defeat the embargo, including false destinations, switching destination at sea, blind shipments, losing oil, and many others. They tried them all with some success.

The market price, standing at about $3.65 a barrel during the summer, began to climb the moment the cutbacks began. Independent refiners and oil companies that did not supply crude oil sufficient for their refineries bid higher and higher, buying whatever and wherever they could to satisfy the demand bloated by the embargo scare. The pressure of consumers, tired of waiting in line at the gasoline pumps, pushed product prices up. Consumer inventories climbed. Every month from November 1973, through the spring of 1974, product prices mounted sharply in most importing countries. The oil companies scrambled for oil.

The governments of oil-importing countries did not cover themselves with glory. If they had backed their companies and had taken a firm stand earlier, OPEC might not have been able to pull off its price coup. Later, at the takeover, the foreign offices of importing countries made no great outcry against the price increase or against

OPEC because it had raised the price unilaterally. Indeed, they indirectly supported OPEC. Even before the price takeover, Dr. M.A. Adelman, highly respected U.S. oil expert, had complained in *Foreign Affairs* (1972) that "OPEC's recent successes must be attributed in large part to misguided policies of the U.S. State Department," a statement he has reiterated many times since. Still, at the critical moment of takeover, events tied the hands of the diplomats.

James Akins, petroleum advisor to the Department of State at the time and later U.S. ambassador to Saudi Arabia, had virtually announced in his article in *Foreign Affairs* in April 1973, that a price of $5.00 a barrel was acceptable. The title of his article, "This Time the Wolf Is Here," indicates its tenor. The article also assured the Arab countries that importing countries, especially the United States, were not prepared for an embargo.

As if to confirm Akins' prediction, when OPEC members raised the price and OAPEC members cut production and mounted the embargo, the United States and other oil importers did not seem to comprehend fully what was happening. Only a year before, both Saudi Arabia and Iran had offered the United States bilateral deals that would have assured its oil supplies. The United States ignored the proposals. Later, American officials could not understand how these two "allies" could damage its interests, the first as the ramrod of the embargo and the second as the leader in pushing the price up.

The embargo and cutbacks not only came as a surprise, but also their degree and effectiveness were even more of a shock. The United States had anticipated a repetition of the farcical 1967 embargo, a token embargo. No policies or actions had prepared the country for a real embargo. When it came, government agencies noisily introduced measures to conserve oil without even knowing the amount of the shortfall. The U.S. government issued a welter of confusing statements and statistics, ranging from a trivial impact to a crippling shortage.

In the crisis atmosphere, officials and citizens thought foolish things. The newspapers had a field day with mixed up, erroneous, and half-true stories. In November, Secretary of State Kissinger through the press ominously warned of "countermeasures." Sheikh Yamani, visiting in Copenhagen, promptly retorted through the press that his country would blow up its oil facilities if the United States ever tried to taken them by force. Ignatius Miles discussed military intervention as a real option in *Harpers* (March 1975). Sometimes it seemed unreal that the United States and the industrial world were being humbled by a clutch of Arab potentates.

United States' domestic production and reserves of crude oil had peaked in 1970 and could not expand. Imports that year provided 22 percent of consumption. By 1973 the United States was importing 36 percent of its needs. As the embargo began to have an impact, the United States frantically sought a way out of the shortage. In retrospect, the United States was not so short of crude oil as it was of refined products, reflecting the misuse of refinery capacity. Indeed, the United States increased its crude imports 32 percent in 1973 and ended the year, despite the embargo and cutbacks, with crude inventories as large as the beginning of the year.

In November 1973, President Nixon made a stirring speech in which he announced Project Independence. That plan would free the United States from dependence on foreign oil by 1980. Knowledgeable people scoffed at the idea and asserted that American dependence on imported oil would grow, not diminish, in the coming years and decades. Conspicuously, the United States retained price controls on oil and oil products, not permitting them to rise to pinch off demand or encourage production. Nor did the United States mount any serious effort to cut consumption, expand oil production, or introduce substitutes. The United States was simply not prepared to take the tough measures necessary to relieve its oil dependence.

In December the Arab position on the cutbacks and embargo began to weaken. The cease-fire held. Sheikh Yamani, after a tour of the industrial countries, feared a deepening recession in the United States, Western Europe, and Japan. He felt that the restrictive measures had served their purpose in shocking the oil importers into rethinking their policies. The sheikh told Secretary Kissinger that Saudi Arabia would modify its stand if Israel only established a time-table and started its withdrawal from Arab lands. On Christmas Day, OAPEC announced an increase of 10 percent in crude production instead of the 5-percent cutback decreed earlier. Sheikh Yamani now said, "We only intended to attract attention to the injustice that befell the Arabs."

Oil production increased in December. For all OPEC members, the increase was one percent over November, about the same as for the embargoing countries. Saudi Arabian production increased five percent. Since production had expanded rapidly in the first three quarters, production for the entire year was well ahead of 1972. Exports of all OPEC members in 1973 were 16 percent greater than in 1972. Saudi Arabian production was up 25 percent for the year. Even with the cutbacks, 1973 was world oil's banner year.

The world oil market was in turmoil. No one quite knew what was happening. The uncertainty and shortfall of exports during the last three months of 1973 propelled the market price upward. American independents and European and Japanese firms bid higher and higher for crude oil, fearful that every barrel would be their last. The companies and OPEC met in November to discuss—not negotiate— the price of oil and the state of the market. Neither side could make any sense out of its wild gyrations.

THE PRICE EXPLOSION

The market was out of control. Neither OPEC members nor the companies exercised much influence so long as such great uncertainty about supply and demand reigned. Some members of OPEC raised their own tax-reference, buy-back, and posted prices weekly, trying to shadow the elusive market price. The posted price of the Persian Gulf OPEC members remained officially at $5.12 a barrel and exporters honored most contracts at the price. The open market price, by everyone's guess, exceeded the posted price substantially by early November. OPEC's intent on October 16 had been to set a posted price well above the market price. By mid-November it was clear that OPEC would soon set another and still higher price.

In November Nigeria offered crude oil at auction, not knowing what it would fetch. It brought $16 a barrel. Other countries set dizzying prices or held auctions that resulted in bids from $15 to $20 a barrel. In early December the National Iranian Oil Company announced an auction for deliveries in 1974 and received bids of $17 a barrel. Libya auctioned its premium light oil at $20 a barrel. The market was riding high and rising. Its price signals reflected ignorance and doubts about supplies—the unknown effects of the cutbacks—not normal supplies.

The OPEC oil ministers met in Teheran on December 22 to reconsider the posted price. In the pattern to be repeated many times later, two positions emerged, represented by the Shah of Iran and Sheikh Yamani of Saudi Arabia. Iran and some Arab countries, eager for more revenues, wanted to set a high price, at least $14 a barrel, reflecting the bids of November and December. The shah wanted to exploit to the fullest the OPEC advantage. He even argued that the price should really be much higher, just a hair below the cost of oil's substitutes. Although it was the Arab cutbacks that had created the supply conditions that made a large price hike possible, it was the Shah of Iran who was most anxious to wring the last dollar out of those special market conditions.

Saudi Arabia, already swimming in money, fought to keep the price increase down. Sheikh Yamani did not want to rupture his country's good relations with the United States. He also feared that a high price would damage the economies of the oil-importing countries. A high price might even reduce exports enough to cut the revenues of OPEC members by stalling the growth of oil importers and inducing a recession. Sheikh Yamani also argued that the auction bids and existing market price were inflated by the cutbacks and that when normal supplies began to appear on the market, the price would fall sharply. He wanted $7.50 a barrel.

The compromise was closer to what the Shah of Iran wanted than it was to Sheikh Yamani's preference. The sheikh believed that the compromise price was too high. As always, he sought the advice of King Faisal. This time the king was not available. Sheikh Yamani had to make the decision alone. The problem was simple: go it alone with a lower price for Saudi Arabian oil with the consequent damage to OPEC and perhaps even the destruction of its still fragile market control, or, for the sake of OPEC unity and in order to consolidate its new market power, accept the higher price. He accepted, but not without agony. Later, King Faisal reprimanded him. But the king ratified the decision.

The Shah of Iran had prevailed despite the pressure for restraint from Saudi Arabia. The shah announced the new OPEC price of $11.65 a barrel on December 23, even before the oil ministers had concluded their meeting. The new price was well below the recent bids, reflecting the caution of Saudi Arabia. But mainly, it reflected both the daring and the vindictiveness of the Shah of Iran.

The new price also eliminated the difference between the posted price and the market price. A posted price continued to govern country-company tax, royalty, and sharing arrangements. But the new Teheran price was the first OPEC price, the minimum price at which OPEC members would sell their crude oil to the oil companies and refiners either by contract or on the open market. When it was announced, all members brought their prices into line with it. The market fever broke.

Quietly and unnoticed, the new dominance of OPEC passed its first test in that December meeting. Unless Iran and Saudi Arabia stood together on the new OPEC price, the frail beginnings of OPEC market control would have disintegrated. The new sense of power of OPEC was reflected in the statement of the shah when he announced the price. He pointed out that the price was low by comparison with the auction prices—two days earlier Nigeria had received bids of $23 a barrel. He argued that the oil companies were still making excessive profits and that consumers were still getting a bargain compared to substitutes.

The Shah of Iran scolded the oil consumers. He said, "the industrial world will have to realize that the end of the era of their terrific progress and even more terrific income and wealth based on cheap oil is finished. All those children of well-to-do families who have plenty to eat at every meal, who have their own cars, and who act almost as terrorists and throw bombs here and there, will have to rethink all the privileges of the advanced industrial world. And they will have to work harder" (as reported in the *Middle East Economic Survey*, December 28, 1973). The oft-lectured was lecturing.

The world stood aghast. The new price was 2.3 times greater than the October 16 price, 3.9 times the October 15 price, 4.4 times the January, 1973 price, and 6.5 times the price of the middle of 1970. Even as late as 1972 in his classic book, *The World Petroleum Market* (Resources for the Future, Johns Hopkins University Press, 1972, p. 258), M.A. Adelman said that "they (producing countries) will inevitably chisel and bring prices down," and foresaw an "inevitable long decline in prices" (p. 262). Most observers saw little to disturb the tranquil oil market of the late 1960s and did not expect major changes in the 1970s. They had not read aright the trends of production, consumption, and reserves. Nor had they foreseen the successful OPEC power grab.

The embargo gradually faded away. Saudi Arabia did not officially declare its embargo ended until March 18, 1974. But in the January-March period, OPEC countries were producing almost at the level of the previous October. Saudi Arabian production was even higher. Arab countries did not enforce the embargo. Oil shipments resumed and inventories everywhere regained their normal levels. The lines at the pumps disappeared. The immediate crisis was over. The long-term crisis was just beginning.

The Arab cutbacks and embargo did not achieve their goals. The cease-fire negotiated with American help ended the fighting, but not the war. It was a small victory for the Arabs. They regained a tiny portion of their lost land. They proved that they were not impotent at arms. More important to the Arabs, and attributable more to the oil weapon than to military action, the policies of Europe and Japan cooled toward Israel. These countries decided that their relations with Israel were less important than were their relations with countries that could assure the continued flow of oil.

The pro-Israel tilt of the United States persisted. There was a subtle change in emphasis. A new urgency in finding a solution to the Middle East dilemma emerged. The United States increasingly played the role of honest broker, trying to arrange an acceptable settlement without favoring either side. The U.S. official attitude toward the Arab countries improved and American pressure on Israel to resolve its conflict with the Arab countries mounted. The United States became the Middle Eastern mediator.

THE EFFECTS

During and after the crisis, the oil companies received a thorough drubbing by the press, people, and governments of oil-importing countries. They were blamed for profiteering, not handling the OPEC members on price matters more effectively, withholding information, causing the oil shortage, not breaking the embargo, and serving the interests of Arab countries rather than those of their home countries. The press snidely called them the "tax collectors" for the "contemptible cartel."

During the crisis the oil companies announced their 1973 profits. They were up 80 percent for Exxon, over 90 percent for Gulf, and high for all of them. The profits of the leading thirty oil companies for all of 1973 jumped 71 percent over the previous year. Rational explanations—unusually low profits the year before, valuing inventories at higher prices, the unexpected run-up of product prices—did not diffuse the public's indignation. Congress scheduled investigations and called the companies on the carpet. At the end of lengthy hearings in 1974 the companies were nominally exonerated of any wrong-doing. But the publicity, scoldings by congressmen, and selective press coverage smeared the companies' images.

The companies were almost as surprised by the events as were governments and the public. Although aware that change was afoot in world oil, they did not expect to lose the power to set prices nor to witness a four-fold price increase in two and one-half months. Although aware that war was coming and that the Arab countries would use the oil weapon, they did not expect those actions to be effective, an expectation shared by governments. They had expanded production rapidly in the first nine months of 1973 as a hedge and had later assisted in softening the effects of the embargo. Still, as appropriate to large international enterprises, they served primarily their own interests.

The interests of the companies changed when OPEC members captured the price and started to control production. So long as the companies controlled the crude oil and the entire chain of operations from desert well to gasoline pump, the companies focussed on the market for final products, serving the interests of consumers with low prices, and of course, their own goals of profits, growth, and stability. With the chain of operations severed into two pieces, with OPEC in charge of upstream activities and the companies running downstream operations, the interests of the companies changed. Like most enterprises, they now have to buy in one market and sell in another.

As long as OPEC members maintained their tight grip on crude oil, the companies found it advantageous to get along with them, accepting their price and production decisions. Otherwise, they would have no oil with which to continue to make money from the part of the chain that they still controlled. With markets for final products secure and prosperous, OPEC members and the oil companies joined in extracting an extra payment—for OPEC—from consumers. To the companies it was not the best solution. They would have preferred to continue their control of crude oil and not to make the extra payments to OPEC. But they didn't and couldn't control crude. Better to share the benefits with OPEC than to have no crude, no products, and no benefits.

The companies, closer at hand to irate consumers, took more heat than did OPEC. For consumers and politicians, however, OPEC also became the enemy. Many thought that it was an OPEC rather than an OAPEC embargo. Some thought that OPEC was an Arab organization and even that Iran was an Arab country. Some wanted to hail OPEC into U.S. courts on antitrust charges. The United States legislated trade restrictions against all OPEC members, even those that increased oil deliveries to the U.S. during the embargo. The price action, embargo, cutbacks, and the war intermingled in the public mind and all were laid at the doorstep of the companies, the Arabs, and OPEC. The quarreling sheikhdoms and banana republics became a malevolent cartel.

The events of the fall of 1973 hastened the reevaluation of world oil resources. Always before, the oil-importing countries had just assumed that the oil was there and that it was theirs, no matter what amount they needed. The export shortfall and price increases of late 1973 forced an examination of rates of consumption, production, and imports, as well as oil reserves. Researchers set to work on new supply and demand studies and projections into the future. The results were shocking. The new calculations spelled doomsday for oil.

The discovery that oil production in the United States and its oil reserves were declining astounded leaders as well as the public. The United States had only about 35 billion barrels of oil left, enough for only a little more than a decade. The industry and the experts had cried wolf too many times. Now that the wolf was huffing and puffing, ready to blow the door down, nobody wanted to believe them. People had difficulty accepting that the world oil scarcity decreed permanently high and rising prices. Unwilling to accept the blame themselves, consumers searched for scapegoats— the companies, government, Arabs, OPEC, even Israel. Although no coherent oil policy emerged from the crisis, oil suddenly became the high-priority item in government and public discussion.

The full bill to pay for the crude oil at the new price would not fall due until later. But in 1973 it was clear that it would be astronomical. Money to pay for the groceries in St. Louis and Bremen went instead for Cadillacs for Kuwaitis, jets for the shah, new factories for King Faisal, and more housing in Caracas. Saudi Arabian marker crude earned for the country $1.77 a barrel on October 1, 1973, up from $0.99 on January 1, 1971. After October 16, the government take was $3.05 a barrel. Following the December price increase, Saudi Arabia received $9.27 a barrel. By contrast, the cost of production of its crude oil was only about 16 cents a barrel.

The world faced a financial panic. Billions of dollars shunted from the coffers of oil importers to the treasuries of OPEC members. Many countries had to borrow to pay their oil bill. Deficits of importers soared and their surpluses and reserves vanished. The less-developed countries peered into the abyss of economic collapse. Saudi Arabia and other OPEC members, also stunned by the deluge, didn't know quite what to do with the money. Frustrated bankers shoveled billions from one bin to another. The world monetary system teetered.

The cutbacks and embargo had effected less a permanent change than created a short period of havoc that left a legacy of uncertainty. The roiling market hastened locking in the higher price and, thus, OPEC control of oil. Despite its weakness, the embargo did change the foreign policies of Europe and Japan and even forced the United States to adopt policies more considerate of Arab countries. More important, the price and production actions combined made the industrial countries recognize their own vulnerability. They scrambled to devise ways to reduce the risk of a repetition of the embargo and to secure independence from imported oil.

The most important event of the crisis was the first. On its own OPEC had increased the price of oil. When the power to decide the price of oil shifted from the companies to the producing countries, substantial price increases became inevitable. The embargo and cutbacks, both resulting from the war, only helped to shorten the time for the appearance of the higher prices. Once the exporting countries decided to control their own production and thus could set the price, they would, given the high rates of consumption and the insensitivity of demand to price changes, inexorably insist upon higher and higher prices.

The price revolution signalled the beginning of the end of the oil era. Cheap oil, based on the payment of production costs and small payments to the owners of the oil, was no more. From October 1973, oil-importing countries would have to pay the owners of the oil what the owners themselves regarded as the appropriate payment. The oil companies, because it was profitable, would help the owners in securing their price. The market, sustained by consumer preferences, technology, the stock of oil-using equipment, and the lack of an alternative, would accept the price and continue to expand.

To understand the full meaning of those critical two and one-half months of late 1973 and what they portend for the future, a trip back through time is necessary. From Drake and Rockefeller, through D'Arcy, the Red Line, the Achnacarry Agreement, the Seven Sisters, and J. Paul Getty, the world oil industry built to a climax in the oil glut of the late 1950s. These events, discussed in the next chapter, culminated in the creation of OPEC.

Chapter 2

From Drake to Getty

Four parties constitute the oil industry. First are the consumers who want to buy oil products. Next are the suppliers, of which there are two kinds: the large international firm that extracts, transports, stores, and refines the crude oil and sells its products to consumers, and the smaller replicas of the larger firms. The third party operates on some part of the vertical ladder from crude oil to consumer. Some operate refineries, buying crude from other firms that do not refine all their own crude. Some produce more crude than they refine. The fourth party of the industry is the owner of the crude oil. In the world oil industry, most of the owners are nations. The governments of these nations are the ultimate suppliers of crude oil.

The development of the oil industry is the story of the jousting for power and position among these four forces. In the very beginning, consumers seemed to have the upper hand. Many small and highly competitive refineries bought crude oil from many small and equally competitive owners of crude oil, all trying vigorously to make their fortunes in the new industry. Soon Rockefeller and Deterding, among others, recognized the potential profits and power of concentrating operations in large units. A few oil giants emerged. A fringe of smaller companies remained but had little influence. Ownership, or at least control through concessions, of the crude oil passed into the hands of large producing companies.

The large companies tried for a period to dominate one another as well as the smaller firms. Their interests—profits, growth, and stability—led them to cooperate with each other, retaining their dominance over consumers and owners and holding the smaller enterprises in check. In the 1950s the smaller firms, by cooperating with the owners, chipped away at the power of the large companies and weakened their influence. In the 1970s, the owners, always before in the backseat of the industry, asserted themselves and dominated the other three participants.

In retrospect, the emergence of the Organization of Petroleum Exporting Countries, the club of the owners, was a natural and predictable stage in the evolution of the industry. The owners always had the ultimate control. But for the first century of the industry, they relinquished their control for a mess of pottage. Small revenues were better than nothing, since the owners did not have the economic or technical resources to develop their own oil. Competition among the owners, fostered by the oil companies, and economic underdevelopment further delayed their progress toward control. Eventually, however, the owners acquired the ability to control their own crude oil. Through the use of their national sovereignty and OPEC cooperation, they captured the industry.

THE FIRST DAYS

Many scholars and historians have chronicled the romance and skullduggery of the industry since Colonel Drake in 1859 drilled that hole in the ground near Titusville. The history of the industry is full of daring, avarice, productive achievement, and the use and abuse of power. Carl Solberg in his *Oil Power* (Mentor, 1976), one of the best of the recent books on the development of the industry, traces the interplay of oil, technology, enterprises, consuming habits, politics, and governments. Anthony Sampson's *Seven Sisters* (Viking, 1975) tells how the giants developed. In the sketch that follows, the focus is on the fluidity of power, its shifts among the components of the industry, and how it has finally come to rest in the hands of the owners.

The price of oil started at $20 a barrel when the well of Edwin Drake, erstwhile railroad conductor and well digger, was producing only thirty barrels a day. That was in 1859. Whale oil was not keeping up with demand and was rising in price. Oil from coal was also expensive. The new substance—kerosene—came just in the nick of time. Within the year the price of crude oil was down to $2 a barrel. During the Civil War, new wells came in, refining sprang up, exports began, and war demands boomed. The price fluctuated between 10 cents a barrel—less than the value of the barrel—and $14 a barrel. The problem was transportation and storage. Oil was cheap just after the wagons arrived and near when supplies dwindled.

Every enterprising soul with a little capital dug a well or built a refinery. One of those testing fortune in the new refining industry was John D. Rockefeller. In 1863 he bought a refinery in Cleveland. He was quick to see that small refineries were inefficient and that competition hurt his profits. He bought up refineries and dismantled many, squelching competition and concentrating production in his most efficient plants. By 1870 he already controlled 10 percent of the American refining industry. It was only the beginning of the first oil empire.

In the confrontation between the refineries and the crude-oil producers, Rockefeller won. After an incipient Pennsylvania producer cartel collapsed, Rockefeller faced many crude producers who competed among themselves to sell him their crude. Rockefeller went on not only to gobble up the refining industry, but also to capture control of the transportation of oil—at first by rail and later by pipeline—and finally to dominate most of the market for oil products. Using practices regarded as ruthless even by the more relaxed standards of the nineteenth century, Rockefeller built a near-monopoly by 1890 and set the pattern for the oil industry of the future.

The key to Rockefeller's success was a large, efficient, and integrated organization. He bought crude oil from a competitive market, refined, transported, and stored it, and sold oil products to consumers. He operated internationally. He produced high-quality products, used the best technology available, and sold at a low price. He was always conscious of the need to keep costs and prices at the minimum. And he suppressed competition. His enterprise was the prototype of the succeeding large oil companies—with one exception. Rockefeller relied on competition among crude-oil owners. He did not own or control his own oil.

THE GIANTS TAKE OVER

Rockefeller's Standard Oil did not endure. Some competition had always existed. The Nobel brothers and Rothschild used their bases in the Russian Baku fields to market oil in Europe and Asia. Marcus Samuels' Shell outgrew selling curios and seashells, also carrying Russian oil to Asia. Royal Dutch began in the Far East. In the United States, the 1901 Spindletop discovery in Texas gave birth to two new companies—Texaco and Gulf. The competition, however, did not bring the Standard empire down. The muckrakers, the trustbusters, and the Supreme Court finally forced the reorganization of Rockefeller's enterprise.

The glare of publicity and militant public policy brought Standard Oil to its knees. Foiled by the courts in his effort to maintain Standard Oil as a trust, Rockefeller reorganized his enterprise by establishing Standard Oil of New Jersey as a holding company to own the other parts of the illegal trust. Bit by bit, Standard's nefarious activities came to light and helped to inspire antitrust legislation. Not until President Theodore Roosevelt's time, however, did the government crack down. In 1911 the Supreme Court issued the dissolution order for Standard of New Jersey, paving the way for the emergence of new Standard companies.

The dismantling of the Rockefeller organization did not end the great concentration of power in oil. Standard of New Jersey, nicknamed Jersey Standard for decades, divested of its offspring, grew and became the world's largest oil company. Today it is known as Exxon. The New York Standard company eventually became Mobil. The third affiliate to join the select Seven Sisters who long dominated the world oil industry was Standard of California (Socal). Others of the original Standard companies have also become powerful oil companies. Rockefeller spawned a family of giants that perpetuated the highly concentrated oil industry.

Most of the Standard companies inherited a deficiency from the crumbling empire. Rockefeller had never interested himself in the production of crude oil. The vulgarity of the roughnecks and roustabouts in the oil fields offended his prim Baptist calling. Instead, he preferred to dominate the competitive producers at arm's length by encouraging more competition among them and through the control of transportation, refining, and the market. The new Standard companies did not control their own crude, sparking an aggressive search for crude in the early part of the century, at first in the United States, then in Mexico, and later abroad.

As early as 1882 Standard Oil set the price of crude oil. The owners of the crude were powerless to do anything about it. Market forces, particularly the condition of inventories, capacity, and reserves influenced the price, and Standard, in setting the price, was careful to observe the market situation. When Texaco and Gulf developed out of the Spindletop discovery at the turn of the century, they accepted Standard's price leadership. When the Court dissolved Standard of New Jersey, the new operating Standard of New Jersey, became the price setter. Jersey Standard (now Exxon) continued its dominance for decades but it always kept a watchful eye on the state of the market, of supplies, and on the activities of other companies.

Two foreign giants emerged to share world oil with the five great American companies. Royal Dutch and Shell, after fighting over five continents, finally joined to become Royal Dutch-Shell under the leadership of Sir Henri Deterding. His was a genuine international company, under British and Dutch ownership. It challenged Standard and later the independent Standard companies everywhere. Royal Dutch-Shell has always considered itself a bit above the American sisters, whom it regarded as rather provincial Yankee upstarts.

Deterding was often in the forefront of new oil gambles. It was he who took the chance on Venezuelan oil. Indeed, he had the lion's share of Venezuelan oil until Exxon, with the help of American diplomacy and chicanery, muscled in. He also developed the Mexican and Romanian fields, often just a step ahead of the Americans. Royal Dutch-Shell even won a toehold in the American market, much to the chagrin of Exxon and other Standard companies.

Although the British had an interest in Royal Dutch-Shell, the British government wanted a Middle Eastern supply that it controlled directly in order to guarantee its naval supplies. The D'Arcy concession had borne fruit in Persia (Iran), and in 1914 the Anglo-Persian Petroleum Company began to produce the oil that powered the British fleet. Six years later the British government became half-owner of the company. Another sister, British Petroleum, so named in 1954, was born.

By 1920 the seven companies dominated the oil industry worldwide. Three were offshoots of Standard—Exxon, Mobil, and Socal. They had production and markets at home and abroad but were clearly American companies. Two more—Texaco and Gulf—derived their rapid growth from Texas and Oklahoma oil fields, but were beginning to put out feelers overseas. British Petroleum owned Iranian oil. The last, although not the least, was Royal Dutch-Shell, operating everywhere. Some lesser companies provided the competitive fringe of the industry and complicated the lives of the great companies.

The giants were vertically integrated companies from spudding in above a promising formation to delivering the product to the consumer's doorstep. Not all of them were balanced companies. Some had more crude than they could refine. Others had a larger market for products than they had crude oil or refinery capacity. Independent crude oil and wholesale product markets developed, providing opportunities for many small companies operating on only a part of the vertical ladder. The oil business originally supplied kerosene, but that product had long since faded into a long list of products. By 1920 gasoline had taken over as the main product as the world fell in love with a motor on four wheels.

The dominance of the industry by the seven companies did not mean that they acted in unison or that they always engaged in cut-throat competition. On occasion, when their interests clearly coincided, they seemed to act as one. On other occasions, competition among them was bitter. They could always be counted on to pressure and, whenever possible, to exterminate the small companies. Among themselves, they penetrated and built markets on one another's turf and stole, by fair means or foul, one another's supplies of crude oil. Oil was not a genteel business.

Price cutting, bribery, and political influence peddling, along with monopoly pricing and exploitation where one company was in control, propelled the industry forward, lurching and stumbling. The business practices of the large companies often served to kill off potential competition from independent companies before they got started. Although never able to eliminate the independent companies completely, the large companies kept them in their place and kept them small enough only to be troublesome. Open warfare among the giants punctuated periods of calm.

THE ACHNACARRY AGREEMENT

One of the fiercest rivalries was between Walter Teagle of Exxon and Sir Henri Deterding of Royal Dutch-Shell. They had met early, back in 1907, when Teagle was still a Standard underling. Over the years they had grown to respect and fear one another. After the Russian revolution, Teagle, in one of those colossal mistakes that oil men make from time to time, bought into the Russian Baku fields, expecting that it would revert to private ownership. Deterding, needing crude oil also, wisely bought it from the new Communist owners. Then he undersold Exxon in Europe and Mobil in the Far East with the oil that Teagle regarded as properly his.

Teagle and Deterding met in 1922, trying to restore order to the chaotic world oil market. The two companies were undermining one another in obtaining Russian crude and cutting each other's throat in the Indian market. British Petroleum was also troublesome in opening its Iraqi fields. Mexican production was expanding chaotically. Crude production was outstripping product sales. Price wars and cutthroat competition created instability for all the large companies. The 1922 effort at cooperation failed. By 1927 desperation forced the leaders to try again to contain the growing disorder of the continuing glut. In 1928 Exxon, Royal Dutch-Shell, and British Petroleum started negotiating.

Achnacarry Castle was the scene. Sir Henri Deterding had rented the Scottish estate for the season. Grouse abounded and the streams promised good fishing. Deterding enticed his guests with happy hunting and new oil industry. In greatest secrecy they arrived, one by one—Teagle and his entourage, Sir John Cadman, new president of British Petroleum, and his assistants, and the leading lights of lesser companies. They did fish and hunt, but not very much and not very well. They did much better at organizing the oil industry and creating a cartel. Each company took home trophies called stability, profits, and the growth of the company.

The central ingredient of the Achnacarry Agreement was the decision that each company would retain its existing markets and crude sources. It guaranteed the relative position of each company. So long as a company made no inroads into the sources or markets of others, it need fear no incursions by other companies. Oil men later called it the "as is" agreement. (The arrangements were revealed in 1952 in *The International Petroleum Cartel*, Federal Trade Commission, 1952.) The most important of the seven rules of the agreement was "The acceptance by the units (companies) of their present volume of business and their proportion of any further increase in production."

No more predatory market penetration or raw swindles to get more crude oil would mar the industry. Each company would restrain itself to its own growing markets, with a guarantee that "only such facilities to be added as are necessary to supply the public with its increased requirements of petroleum in the most efficient manner." Efficiency had always rated high with the oil companies. By definition, competition reduced efficiency and must be eliminated.

The companies chose a new way to set the crude price—Gulf plus. The Gulf of Mexico price plus transportation costs from the Gulf protected the higher cost American companies and guaranteed profits for all companies. The price delivered at any point in the world was the Gulf price, dominated still by Exxon, plus the transportation charges from the Gulf to that point, regardless of the actual point of shipment. Crude delivered to Spain from Iraq paid the Gulf price and what it would have cost to get the oil from the Gulf to Barcelona.

When the companies swapped crude at different locations, a common practice, each pocketed one-half of the gain from shipping the crude the shorter distance. Rivalry for Mexican crude was stopped by dividing the country up among the companies. Exxon and others received a slice of Iraq. The agreement had something for everyone. All the American sisters joined Exxon, Royal Dutch-Shell, and British Petroleum in creating the cartel.

The agreement was possible because Exxon and Royal Dutch-Shell, the two largest companies, finally recognized that neither could make Rockefeller's dream of world oil monopoly come true. Each checked every move by the other. Neither could gain ground on the other without a grave risk of losing ground. The next best thing to a monopoly, the companies reasoned, was a cartel that assured a stable price and a guaranteed share of growth without the hazards of inefficient and disturbing competition.

The cartel agreement gave birth to many lesser agreements among pairs of companies and groups. These agreements divided up crude supplies and markets, mostly in accordance with the "as is" spirit. Neither the main agreement nor the corollary agreements worked perfectly. Despite formal and written undertakings, some companies cheated on the rules. The declining markets of the Great Depression and later the development of new fields, such as those in Kuwait and Saudi Arabia, were difficult to handle. All companies eagerly sought the new crude and markets, but now with some rules to the game.

As an agreement binding on its signatories, as well as a statement of principles and intentions, the Achnacarry Agreement modified the behavior of the companies enough to ameliorate the glut and restore the stability of the industry. With the return of stability came ample profits and steady growth. What no single company could have achieved alone, the companies in cooperation could and did achieve. Later, the oil-exporting countries, in seeking to increase their benefits from oil, remembered that cooperation among the companies had yielded great benefits.

The next big glut originated in the American rule of capture. Under it, oil belonged to those who found it and pulled it out of the ground first. The large companies were powerless to stem the new tide of crude from the East Texas boom of the 1930s. Every wildcatter in the country who could borrow the money rushed to Texas to get his share of the bonanza. He would never agree to cut his production to benefit the driller on the next hillock.

Agreement finally came, but not from the producers. It came from the outside. The State of Texas, alarmed at the flood of crude and prodded by those who would benefit most—the large companies—set up the Texas Railroad Commission by curtailing production and stabilizing the market. Well by well and company by company, the commission prorated and regulated production until the crude price recovered, and order in the market once again ruled. The oil-exporting countries also learned well the lesson of the Texas Railroad Commission.

MEXICO NATIONALIZES
AND FADES

The first battle between the oil companies and an oil-producing country came in Mexico. Its expropriation of the foreign oil companies demonstrated their vulnerability. For the first time, the companies learned that if their behavior transcended some limits, such as defying national authority, a nation could and would strike back. Although nationalization killed Mexico's export trade in oil, it also showed that a nation could survive the wrath of the oil companies. Oil-exporting countries learned that they were not without power and that oil companies did not possess limitless power.

Mexican oil development began shortly after the big Texas field came in at the turn of the century. Two entrepreneurs, one English, the other American, fought up and down the east coast of Mexico for oil. Royal Dutch-Shell, Exxon, and others joined later. Edward Doheny, later of Teapot Dome infamy, bought land that included mineral rights—an exception to the usual Latin American system of reserved subsoil rights—and brought in the first well in 1901. Soon his strip of land near Tampico was called the Golden Lane.

Not to be outdone, Weetman Pearson, a Yorkshireman who later became the first Lord Cowdray, obtained concessions, struck oil, and founded El Aguila. By 1910 Mexico was the second largest oil producer in the world. Before long the big companies decided that Mexico would one day rival the United States and many hurried to stake out claims. Royal Dutch-Shell bought into the Cowdray interests in 1918. Exxon tried to protect its sources and markets in Mexico and other Standard companies bought old oil fields and brought in new ones.

Rough competition between the British and American companies reduced Mexican politics to shambles. The Mexican revolution created the ideal climate for oil-company manipulation. Although in 1917 a new constitution reinstalled national control of the subsoil, taxes were a pittance and were often uncollected. The companies supported first one and then another petty tyrant, some of whom reached power momentarily. Production grew, as did profits. Mexican oil played an important role in the first world war.

Mexico was a part of the instability that the Achnacarry Agreement sought to resolve. After the accord, the companies treated one another more circumspectly. But the agreement made the companies complacent about the treatment of the producing countries. If the other oil companies agreed not to undermine them, then certainly no dozing campesino propping up an adobe hut or a cross-bandoleered Mexican general could challenge the companies.

The expropriation began as a labor dispute. The companies paid their workers miserably. American cattle lived better. The Mexican labor union demanded higher wages in 1936. The companies, never stooping to bargain with the union, offered one-fifth of what the unions wanted, pleading that they could not afford more. In May 1937, the workers struck. They placed their case before the Federal Board of Conciliation and Arbitration. The board hired experts who, not privy to company information, estimated that the companies could pay double what they had offered. In December 1937, the Federal Labor Board ordered the companies to pay. The companies appealed to the Mexican Supreme Court. The Court affirmed the board's decision.

The companies then made their fatal move. They defied the Mexican Supreme Court, saying simply that they wouldn't pay. Lazaro Cardenas, the Mexican president, could not tolerate this open defiance of Mexican sovereignty. In May 1938, he expropriated the seventeen foreign companies and created Petroleos Mexicanos (Pemex) to take over their properties. The oil companies were stunned. They felt it could not be done. When the shock wore off, they proceeded to scrub Mexico from the world oil map.

The companies appealed first to their home countries to send in troops to protect their fields. President Franklin Roosevelt, determined to live up to his Good Neighbor policy, turned them away with a lecture on how the rich should treat the poor. The Monroe Doctrine protected Mexico from the British. Then the companies had to fall back on their own resources. They organized a boycott and pressured most of the potential buyers of Mexican crude into accepting it. In an uneven fight, Mexico lost.

The Mexicans did continue to produce oil. They even exported some of it after a while as war enveloped Europe and Mexico threatened to sell oil to Germany. Even the United States bought some Mexican oil during the war. Eventually, the companies lost interest in punishing Mexico. But Mexico did not enter the export market as a major force again for decades. Instead, Pemex developed its domestic market and over the years has aided substantially in the industrialization of the country. The expropriation hurt, but Mexico survived, even if it wasn't on the oil maps.

The companies regarded it as a clear-cut victory. Mexico eventually had to indemnify the companies. The boycott eliminated Mexico as an exporter of oil. The companies believed that they had taught Mexico, and the world, the hard lesson that international oil companies were sacred, untouchable by producing countries, except at their peril. No country, they felt, would ever try it again.

But ever after the Mexican expropriation, producing nations—and the new ones coming along—toyed with the possibility of nationalization. It was something that could be done *in extremis* now that Mexico had pioneered the way. Companies' haunting fears of it also strengthened the oil-producing countries and brought the companies to the bargaining table more readily. The bitterness of the Mexican case and losses to Mexico propelled producing countries to the view that although one nation cannot win against the oil giants, perhaps a group of nations could one day challenge them.

VENEZUELA AND THE 50-50

The world barely missed the absence of Mexican oil because a new large producer was rapidly expanding—Venezuela. The Indians had used oil as a medicine and the conquistadores exported two barrels of oil in 1539 to cure the Spanish king's gout. Pirates later caulked their ships with heavy oils that seeped from the ground. But the first economic use of Venezuelan oil did not come until 1878 when Manuel Pulido dug a well by hand, refined his crude, and peddled kerosene house to house up in the Andes Mountains.

Caribbean Petroleum brought in Zumaque-1 in 1912. It was the first really commercial well. Deterding of Royal Dutch-Shell bought Caribbean Petroleum in a speculative venture even before any large oil fields were found. The first great oil hunt was on. The speculation paid off handsomely. Exports began in 1917, and within a decade Venezuela was a major exporter, supplying the American and British markets. The British and Dutch at first captured most of the good concessions. After the first world war, American companies obtained a foothold. Exxon soon began to rival Royal Dutch-Shell. The oil companies and the Venezuelan military shared the domination of the hapless country. My book, *Venezuelan Economic Development* (JAI Press, 1977), traces the development of Venezuelan oil and oil policy.

General Juan Vicente Gomez, dictator from 1908 to 1935, ruled Venezuela with a fist of steel, with prisons that only Hitler would later rival, and a spy system that the FBI and CIA would envy. He treated the country as his personal ranch. He trafficked in concessions, condoned corruption and criminality, enriched himself and his family, and kept taxes on the oil companies low. The oil companies wrote the petroleum laws and arranged to have ministers appointed and dismissed. The morals of the companies were about the same as those of the bachelor general who fathered a hundred children. He died the wealthiest man in the country, owner of nearly two-thirds of the arable land. The companies prospered and the oil came gushing out.

The oil companies earned the antipathy of the people, as they had in Mexico. When General Gomez died, the cry soon went up to nationalize, as Mexico had. But Mexico was a big country and had other products. Venezuela had oil and a few coffee trees. The successors of General Gomez, Generals Eleazar Lopez Contreras (1939-1940) and Isaias Medina Angarita (1941-1945), did gradually tighten the screws on the high stepping countries, under mounting public pressure. Antioil company novels became an art from, and no Venezuelan had a good word for the companies. General Lopez Contreras raised taxes a bit and permitted labor unions. Modest regulation of the industry began. General Medina Angarita went much further.

Venezuelan petroleum laws were a mess. Concessions had been granted under a dozen different laws, each calling for different royalties, time periods, and taxes. Complex and shady trading in concessions had clouded the legality of many of them. Rising popular sentiment continued to protest the low taxes and the failure of oil companies to obey the laws. The companies paid no duties on their imports, for example. General Medina Angarita determined to reform the whole industry.

In 1943 Venezuela passed an income tax law, the first among the oil-producing countries. Ostensibly for everyone, only the oil companies paid income taxes in those days. And they were steep. The goal of the law was to capture, in royalties and taxes combined, one-half of the oil companies' net income. The law didn't quite produce that much, but it came close. Venezuela was the first to test the companies on taxes and to test the provisions of U.S. tax laws that permitted the companies to use Venezuelan taxes to offset U.S. taxes.

The next step, taken in the same year, was the new petroleum law. It was important not only because it governed the industry for the next thirty-two years, but also because it set the tone of relations between the companies and the government. It was more than just a law. It was an agreement achieved through bargaining between the companies and the government. From 1943 until nationalization in 1976, the Venezuelan government attained most of its goals through negotiation.

The new law cancelled all existing concessions and granted new ones to the same companies covering nearly the same areas. All the concessions were for forty years and carried higher royalties—16 2/3 percent—than the old royalties. The law wiped all legal stains from the concessions. It introduced new regulatory and conservation measures. It required the oil companies to build their refinery industry in the country. The oil industry was declared a public utility that would revert in its entirety to the state in 1983 without compensation. The agreement and law were a credit to both the oil companies and Venezuela.

The companies were happy with the new law because of the new legal concessions for the long period. They promptly expanded and made Venezuela the number-two producer and number-one exporter in the world. The income taxes did not bother the companies much because they simply offset their tax obligations in their home countries. Venezuelans were happy with the greater taxes, the more effective control of the industry, and the assertion of their national authority. Cries of nationalization died out but the general's political opposition voted against the law. The new democratic forces regarded it as too soft on the companies. Among the young politicians voting nay on the new law was Dr. Juan Pablo Perez Alfonzo, the founding genius of OPEC.

Government pressure did not end. In 1945 a new democratic reform government elevated Dr. Perez Alfonzo to oil minister. He insisted on a special tax that delivered 50 percent of company profits to the government, the famous 50-50 profit split that spread like wildfire through the Middle East. He also introduced a policy of granting no new concessions and was readying legislation to create a national oil company. On the domestic political scene, however, the new government moved too fast on too many reforms. In 1948, the military recaptured the government.

The new military government left untouched most of the petroleum reforms of the democratic government. Its interpretation of the laws, however, was more generous to the oil companies. When the companies complained of dwindling reserves, General Marcos Perez Jimenez opened the country in 1956 and 1957 to new concessions. The independents were the pushiest and received many of the new concessions, weakening the position of Royal Dutch-Shell, Exxon, Gulf, and Mobil.

Despite the Suez crisis the petroleum market in 1957 began to soften. Just as the new concessions started to produce, the glut appeared in the market and the market price started to decline. In 1958 the dictator departed hastily for Miami, and in 1959 the democratic government returned. In the same year, the oil companies cut the posted crude-oil price. To bolster revenue, the government raised the ante to two-thirds of oil profit for it, one-third for the companies. Dr. Perez Alfonzo was back at his post, now calling loudly for international cooperation.

THE RED LINE

When the first world war ended, the Ottoman Empire, along with the German Empire, died. The victors hungrily pawed over the carcasses. The presence of oil in the Middle East was, of course, well known. British Petroleum already had the concession covering Iran. Before the war British Petroleum had also acquired almost half of the Turkish Petroleum Company, along with Royal Dutch-Shell and the German Deutsche Bank as partners. This enterprise had been put together by one of those pioneering entrepreneurs who lent intrigue to oil history, Calouste Gulbenkian. He wisely kept 5 percent of the concession for himself. The company was dormant during the war.

In 1919 the British and French carved up the Ottoman Empire into mandates. Arab nationalism and Lawrence of Arabia were all right to help win the war, but not for the peacetime control of oil. The French took the German share of Turkish Petroleum. The Americans were left out. They hadn't declared war on Turkey. American officialdom banged on the Middle Eastern door long and hard, demanding an open door, for Americans at least. The British and French finally opened it just a crack, letting a few Americans in, but then they all slammed it shut again.

The new Americans in the Middle East were Exxon, Gulf, Texaco, Mobil, and three other companies. They initially acquired 20 percent of the Turkish Petroleum concession. The American Middle Easterners did little to exploit their opportunity at first. Soon the new country of Iraq signed an agreement with the oil companies, creating the Iraq Petroleum Company as the successor to Turkish Petroleum.

Since Iraq Petroleum had originated with Gulbenkian's creation before the war, that savvy oil man had insisted on a stipulation. The participants must agree not to seek concessions anywhere in the Ottoman Empire except through the new Iraq Petroleum Company, of which he owned one-twentieth. But no one knew the limits of the Empire. So Gulbenkian took a map and drew a line with a red pencil around an area that included present-day Turkey, Syria, Jordan, Israel, Iraq, and Saudi Arabia. This line did not enclose Kuwait or the offshore islands.

This was the famous Red Line Agreement, an accord that ultimately delivered Middle Eastern oil to America. Involved in the agreement were British Petroleum, Royal Dutch-Shell, the Compaignie Francaise de Petroles, and five American companies, including Exxon, Gulf, and Mobil. These companies all opted to limit themselves in the Middle East—except in Kuwait, the offshore islands, and Iran—to Iraq Petroleum and the concessions it might obtain. Except for the British Petroleum concession in Iran and the Iraq Petroleum concession in Iraq, just getting started, the Middle East in 1930 remained virginal.

Another of those enterprising loners of oil, Frank Holmes, now enters the picture. Outside the red line, the beginning of oil in the Middle East was in Bahrain, islands off Saudi Arabia in the Persian Gulf. In the mid-1920s, Holmes had secured a concession from the sheikh whose land was still under British protection. Britain, with Iran in its pocket and the control of Iraq Petroleum, lacked interest in tiny faraway islands. Holmes then went to the United States to peddle his concessions. Teagle of Exxon, also uninterested, committed the monumental error of turning it down. Gulf bought an option on the concession and explored the area. But even though the indications were favorable, the Red Line Agreement, which Gulf had just initialed, precluded further Gulf interest.

Gulf offered the option to Socal, not a party to the Red Line Agreement. Over British objections and only through a Canadian subsidiary, Socal obtained the concession in 1930, struck oil in 1931, and by 1933 was exporting oil. Bahrain never amounted to much as an oil producer. It has become a major banking and trading center, renewing its ancient role in Middle East commerce.

Bahrain's role is important in Middle Eastern oil because it inspired King Ibn Saud of Saudi Arabia to wonder if his vast desert secreted oil. Karl Twitchell, an American geologist, studied the area and told the King that oil was there, despite formations different from those in Iran and Iraq. Twitchell brought the proposal from the king to the American companies. Most of the American companies, faced with a glut occasioned by the Great Depression, and in any case bound by the Red Line Agreement, did nothing.

Caught napping in Bahrain, British Petroleum came to life at the prospect of Saudi Arabian oil. The offer of Iraq Petroleum on behalf of British Petroleum, however, was paltry—rupees and not enough. The king wanted gold coins and a lot. After its Bahrain success, Socal decided to take another chance and satisfied the king's fondness for gold. But Socal lacked the capital and markets for a large operation. In 1936 Socal teamed up with Texaco, the only other giant American firm not hemmed in by the red line. In 1939 the oil began to flow out and the gold to flow in. At this point, American companies had penetrated Iraq Petroleum and had buttoned up Saudi Arabia, destined to become the greatest producer of them all.

One major Middle Eastern production center remained—Kuwait. Reenter Frank Holmes. When he sold Gulf the option on Bahrain in 1927, an option on Kuwait went along with it. When Gulf turned Bahrain over to Socal, it kept the Kuwait option because it was outside the red line. British Petroleum, always on the prowl in the Middle East, was not so much interested in producing in Kuwait as it was in keeping Gulf out. After complex negotiations, marked alternatively by amity and treachery, Gulf and British Petroleum joined in 1934 to secure the concession. In 1938 the oil spurted out of the ground and it is still spurting.

Despite the success in Iraq and Kuwait, the real Middle Eastern bonanza was Saudi Arabia. Two American companies—Socal and Texaco—had the concession. Within a few years of the first well, it became obvious that Saudi Arabia had more reserves than the United States, maybe more than anybody, and would soon become one of the world's largest producers. The two companies were uneasy. They lacked the capital for the venture and did not have adequate markets for the crude. They decided to invite other companies to help develop and share their find.

Negotiations began shortly after the Second World War. Socal and Texaco wanted Exxon and Mobil—bound by the Red Line Agreement—to join them. The French and Gulbenkian insisted on maintaining the agreement. The French held out partly as a negotiating chip, hoping to get a slice of Saudi Arabia. Gulbenkian did not want to jeopardize his 5 percent of Iraq Petroleum. Again the State Department came to the aid of American oil. An agreement to expand Iraq Petroleum mollified the French and Gulbenkian.

The Middle Eastern door once again creaked open, only to bang shut again when two Americans had entered. In 1948 the Arabian American Oil Company, holding the world's most precious concession—Saudi Arabia—consisted of Exxon, Socal, and Texaco, each holding 30 percent, and Mobil with 10 percent. The Red Line Agreement died. In its lifetime and through its demise, it had made the large American companies even more powerful. The independents were itching to share some of their profits. The producing countries began to press for more revenues.

Up to this point, the king and the sheikhs of the Middle East had been satisfied with modest payments. They had almost no petroleum policy. Their royalties were 12.5 percent. Their principal interest in oil was simply to obtain some money. But then the nationalism and development mystique of the postwar period overcame them. The presence of Israel prodded their Arab sensibilities. The poverty of their peoples finally sank in.

Saudi Arabia learned of the Venezuelan income tax and of the ploy of Dr. Perez Alfonzo in getting half of the companies' profits. In 1950 Saudi Arabia installed the income tax and demanded 50 percent of profits. Aramco made the payments to the King and charged it to the American taxpayer. It was clever foreign aid, not requiring legislation and more palatable for a pro-Israel America to swallow. The king had more money and Aramco's profits were undisturbed. Soon other Middle Eastern producers got wind of the gimmick and insisted on their share. In just a few years, the 50-50 deal was all but universal. The exception was Iran.

IRAN NATIONALIZES
AND CONSORTS

Iran had been a British preserve dating back to before the first world war. It was not a colonial outpost, but rather a company country—the domain of Anglo-Persian Petroleum Company. Reza Shah, whose reign began in 1921, restored the splendor of ancient Persia to the monarchy and prospered under British tutelage. But he was not fond of the British. In 1941 the British and the Russians, fearful of German influence when the shah refused to throw the Germans out, invaded Iran and exiled the shah. After the war the British installed his son, Reza Shah Pahlavi as the monarch, expecting the young man to be a willing puppet.

British Petroleum was soon caught up in the 50-50 problem. Venezuela and then country after country in the Middle East had their kitty sweetened at the expense of the United States Treasury. The British didn't want to do it that way. They offered a supplemental agreement that would have given Iran just as much money, perhaps even more. But it was not the 50-50 deal. The magic of 50-50 captured the Iranian National Assembly, the Majlis. It wanted 50-50, not something just as good, or would fight. Nationalistic fever ran high. All things British became unpopular, especially British Petroleum, whose stodgy leadership was the embodiment of capitalist imperialism.

Dr. Mohammed Mossadegh, a member of the Majlis, championed a greater share of oil for Iran. His odd appearance, dress, and personality, as well as his popular cause, made him a national leader quickly. The failure of British Petroleum to move fast enough impelled him to call for its nationalization. He struck a responsive chord. The prime minister, a responsible general, pointed out all the reasons why nationalization could not and should not be done. For thanks, he was murdered. Dr. Mossadegh soon became the prime minister. The Majlis voted to seize British Petroleum properties. Exports halted, the great Abadan refinery shut down, and by October 1951, the last of the British Petroleum employees had sailed for home.

This first challenge to the oil companies since Mexican nationalized oil thirteen years earlier hit British Petroleum the same way the Mexican affront had. It could not be done. But this was more than a confrontation between a company and a country. The British government owned the controlling interest in British Petroleum. The oil men muttered that Iran could not nationalize British government property. But it had.

British Petroleum urged the government to take the only proper course. A few gun boats and a regiment would restore sanity to the weeping Dr. Mossadegh. The British vacillated. Force was too drastic, they feared, for the twentieth century. The company boycotted Iranian oil and lined up all its friends to refuse to buy from Iran. It then sat back and waited for Dr. Mossadegh to collapse. He did not oblige. Without an ally, without exports, without money, Dr. Mossadegh and Iran held out.

Two years later Dr. Mossadegh was still holding out. He appealed to everybody for help, including the local Communists. The Americans lent a polite but deaf ear while they plotted a coup and divined ways to use the situation for the benefit of American oil. Dr. Mossadegh finally stumbled. He assumed control of the army, and the shah tried to fire him as prime minister.

Failing to oust Dr. Mossadegh, the shah fled the country, pleading for help. Violence erupted, and in a plot carefully orchestrated by the American Central Intelligence Agency, Dr. Mossadegh fell. The shah returned, setting the stage for a new order for Iranian oil. The new order necessarily included American companies, since no Iranian government could stand with the British still in control.

The American government was in an awkward position. For years it had patiently gathered information on the big oil companies in order to bring them into court on antitrust charges. In the early 1950s the case was nearly ready. In 1952 the Federal Trade Commission published its study, the work of Dr. John Blair, documenting the existence of the international cartel with the complicity of American companies. The State Department, Pentagon, and the companies wanted a piece of Iranian oil. The Justice Department wanted to break up the companies, not make them stronger. The cry of national security won the day.

Iran refused to return to the concession system. The National Iranian Oil Company owned British Petroleum properties and that was the way it would stay. But it had no markets. The new arrangement, negotiated in late 1953, created a consortium of companies that would buy and market Iran's oil. British Petroleum was the senior member, with 40 percent of the consortium. Five of the large American companies, as well as the French company, were also members. The Americans together had as much as the British. As an afterthought to please the American Department of Justice, nine American independent companies were also cut in for 5 percent.

The benefits to Iran from the consortium were comparable to those of other producing countries. It appeared that Iran had nationalized the foreign companies and had gotten away with it. True, the country had suffered more than two years of stagnation and underwent economic and political turmoil as well as a foreign-inspired coup. And although the oil was Iranian, the large companies were still in control through the consortium. Iran was not aware of a secret deal to limit the growth of Iranian crude production so that no glut would threaten the control of the big companies.

THE RISE OF THE INDEPENDENTS

Although the oil giants dominated the industry, a fringe of smaller companies had always yapped at their heels, injecting competition and uncertainty. They also helped to augment the power of the oil-producing countries by offering them an alternative to the big companies. The larger and more geographically diffuse the industry became, the more opportunities arose for an interloper to get a foothold and to create problems. The small companies, of course, were not small. They were multimillion-dollar sharks swimming in an ocean with multibillion-dollar orca whales.

Most of the independents were American. The old Standard trust and holding company had spawned some of them, including Standard of Indiana (Amoco), Standard of Ohio (Sohio), Continental (Conoco), and Atlantic-Richfield. Others sprang from the efforts of unusual individuals, such as J. Paul Getty (Getty and Tidewater), H.L. Hunt (Hunt International), and Armand Hammer (Occidental).

Still other independents had made it big in the United States but needed to move abroad to secure their own crude oil supplies. National companies of some of the industrial countries also needed crude oil. The Italian and Japanese state oil companies can be counted as independents. The common element in most of these independent enterprises was that they needed crude oil. One independent was different. The Soviet oil-exporting organization had crude oil to sell.

The independents were always there, making money and occasionally irritating the big companies. But they came into their own in the 1950s when they became more than an irritant and began to grab some of the power of the giants. The profits of the large companies were so great that even a small slice of the business offered mammoth fortunes and spurred companies to high risks and great efforts. Some paid off, produced a lot of oil, and tipped the balance of power in the industry.

Getty and a consortium of American independents called Aminoil, which Phillips Petroleum controlled, were among the first to test the water. Concessions were opened in 1948 in the neutral zone between Kuwait and Saudi Arabia. Aminoil got the concession on the Kuwait side of the zone, Getty on the Saudi Arabian side. By offering large concession payments and higher royalties and taxes, the independents beat out the big companies. The king and sheikh welcomed the money and the leverage that the independents provided in dealing with the oil giants.

The Iranian nationalization had offered another opportunity for the independents. When the big companies established their buying consortium in 1954, Washington, greatly concerned about their monopolistic tendencies, forced them to include nine American independents. What the small firms lacked in size, they made up for in political influence. In this case, they used the Justice Department's pending antitrust suit to their advantage.

Libya's new fields also gave the independents access to crude oil. That country, determined not to be dominated by one or a few companies, granted fifty-one concessions to seventeen companies in 1956, most of them to such independents as Continental, Occidental, the Oasis group, and Ameriada. Venezuela had also shunned the big companies when in 1956 and 1957 it granted new concessions.

Enrico Mattei headed the Ente Nazionali Idrocarburi, the Italian state oil company. He was an implacable foe of the Seven Sisters. Indeed, it was Mattei who first called them the *sette sorelles*. His country, on the losing side in the war, had no access to Middle East crude oil. But his country was making rapid industrial progress and required crude oil. Mattei had hoped to join the American, British, and French in the Iranian consortium, but they froze him out.

Mattei retaliated by making what the sisters regarded as outlandish and dangerous offers to the oil-producing countries. In 1957 he offered Iran a 75-25 profit split to develop new oil for the National Iranian Oil Company, but outside the consortium arrangement. Standard of Indiana made similar offers. Mattei enticed other oil exporters with attractive deals, with other independents following in his wake to get a piece of the action. He even made deals with the Soviet oil-exporting organization when the Soviet Union had crude surpluses in the late 1950s. The big companies feared a flood of Soviet oil and a renewal of the glut and price wars that followed the first world war.

Many nations were soon in the act. France, although it had a small slice of the oil market in Iran, was not involved in Saudi Arabia, and aggressively pushed for new oil in Algeria. It brought in large fields in 1956. The Japanese got the Kuwait neutral zone offshore concession in 1957, giving more than half of the profits to the government. Others active in the search for crude, and finding it were Petrofina (Belgium), Gelsenburg (Germany), and Hispanoil (Spain). They offered producing countries better deals than they had received before. Together the independents attenuated the control that the major companies had over crude-oil supplies, fattened the oil-producing countries, and raised their expectations. The giants and independents together now produced more crude oil than the market could absorb.

GLUT AND PRICE

The self-appointed task of the large companies was to maintain order and stability in the industry, assuring growth and profits.

This required manipulating production levels in more than a dozen countries, with each company behaving as a cartel member, even without a cartel. Only in this way could they assure that just enough crude oil came onto the market to keep the price stable or rising gently. When demand slumped, the companies temporarily produced less and let inventories climb. When demand surged, production increased and inventories depleted. The companies shared proportionately in prosperity and depression and in seasonal adjustments.

Production increases in Venezuela, Iran, Iraq, Kuwait, Saudi Arabia, and other countries, including the United States, must conform to the overall estimated rate of increase of the market. When Venezuelan production declined, Iranian production might increase. If an independent charged ahead in Libya, then Exxon had to cut back in Saudi Arabia to preserve the market balance. Obviously, it was not possible to hit the market absorption rate exactly every year, but the giants strove for growth of about 10 percent per year by manipulating demand and orchestrating worldwide production.

The formal cartel established at Achnacarry had long since ceased to exist. By the late 1950s the oil industry was many times the size it was in 1928. Production came from different countries and went to different countries. The companies had learned to regulate their affairs without a formal agreement. Each of the large companies sought the same goals and pursued them by endeavoring to estimate, correctly in most cases, the effects of their actions on others. Even without an agreement, the large companies behaved as though there were an agreement.

During the second world war the United States government controlled prices at about $1.10 a barrel for crude oil. British pressure resulted in a second pricing system, for the Persian Gulf. Its price was about 10 cents less than the United States' price. When price controls went off, both posted and market prices climbed steeply in 1945 and 1946 as the companies tried to restore profits. The American posted price went to $2.50, the Middle East price to $2.20. Then the Middle East price fell back down to about $1.80 a barrel. Into the 1960s the U.S. and Gulf price movements paralleled one another. The U.S. price was constant from 1946 through 1953 and the Gulf price was constant from late 1948 to 1953. Then both prices jumped up. Both recorded another momentary rise in 1956 when the Suez Canal closed.

By 1957 the independents were beginning to make their presence felt, upsetting the tacit agreement among the large companies. India and Italy were buying Soviet oil. The American independents in Kuwait, Saudi Arabia, and other countries were lifting crude oil as fast as they could. The companies marketing oil products were desperately trying to encourage oil's use by offering gadgets and gimmicks. The oil giants were losing control of the industry and instability threatened.

Crude oil began to sell at discounts and then the discounts mounted. The spread between the posted price and the market price widened. The 50-50 deal that had seemed so advantageous less than a decade earlier turned sour and was beginning to hurt. The companies paid royalties and taxes on the fixed posted price, but now received their incomes based on heavily discounted market prices. As the market declined, the percentage of net income going to the producing countries rose and profits declined.

The oil companies sought ways out of this bind. They could not relieve the glut so long as the independents were independent and acted as spoilers. The large companies could cut their own production no more. All of them could indeed produce a lot more than they were producing. They had no control over Soviet exports or over the activities of the growing national oil companies. The only way out was to reduce the posted price which would eventually encourage greater consumption. It would also shift revenues going to governments of oil-producing countries immediately back to the companies.

In February 1959, the major oil companies reduced the posted price about 18 cents a barrel, taking about $150 million from the producing countries. They let out a loud howl. Market prices continued to sag. The estimates of the companies indicated a long period of surplus crude oil and weakening market prices, eating into profits. In August 1960, Exxon cut the posted price again, this time about 5 percent. The other companies followed suit. These price cuts finally stirred the oil-producing countries to action.

Chapter 3

The First Decade of OPEC

Just as a newborn babe, the international organization that saw the light of day first in Bagdad on September 14, 1960, had no power or influence. It had only wants and needs. Like an infant, it kicked and screamed. The oil companies ignored OPEC, hoping it would go away. It would never last, they believed. Arabs would quibble endlessly among themselves and could never get along with non-Arab Moslems, much less Latin Americans. The world did not even know that OPEC existed.

OPEC grew up fast. Ten years after its birth, it was a respected world organization, negotiating in Vienna, its headquarters, with the oil companies on behalf of its ten members who controlled 82 percent of world oil exports. It had achieved modest economic benefits for its members. More important, the OPEC members had learned the oil business and how to control it. They had developed a bold and comprehensive oil policy that if executed would deliver world crude-oil production and prices into their hands.

The decade was OPEC's developmental period. Although as British oil expert Dr. Edith Penrose said in a paper reprinted in *The Growth of Firms, Middle East Oil and Other Essays* (Cass, 1971, p. 236) 1969, "it (OPEC) has helped to clarify the nature of the issue facing companies and governments, to spread understanding of the industry in the oil-producing countries, and to bring the companies to a fuller appreciation of the limitations of their position," the companies still regarded OPEC as a transitory nuisance. To the oil-importing countries, OPEC was a minor problem for the companies to handle. The members of OPEC had faith and hope, but their charity for consumers and the companies was wearing thin.

PRODUCING COUNTRIES
SPEAK UP

The producing countries required no great wisdom to discern that in the 1950s they had no influence in the oil business. The companies had all the chips, including the oil facilities, technology, means of transportation, markets, and vast economic power. The oil-exporting countries competed among themselves to sell more and more of their oil to earn more revenues. That was just the way Rockefeller had planned it for his vulgar Pennsylvania producers in the early days of oil.

All through the 1960s OPEC operated on two fronts. OPEC functioned as an organization and gradually established itself as the voice of the oil-exporting countries. In addition, each of the members continued to negotiate with its own companies on a broad range of topics, including royalties, taxes, regulation, concessions, prices, discounts, and many others. Over a series of years, OPEC has published volumes entitled *Selected Documents of the International Petroleum Industry*, detailing these agreements, laws, and policies.

OPEC as an organization, that is, the oil ministers of all members in conference and their secretariat, supported the moves of each member. OPEC *Annual Reports* give the resolutions and activities of each conference. In the *OPEC Bulletin* (Vol. 9, No. 33, August 1978) appears a summary chronology. Fuad Rouhani, the first secretary general of OPEC, in *A History of OPEC* (Praeger, 1969), shines the light of an insider into the complex negotiations of companies and countries. Out of the tangled web of circumstances and events in the 1960s emerges the story of petty, intricate, and almost endless negotiations. The results were small economic benefits for members and the establishment of OPEC principles that gained clarity and strength through the decade.

The companies had contracts, concessions, and legal arrangements accepted by inexperienced and immature governments that legitimized their position and rights. The oil-producing countries had learned from Mexico in 1938 and Iran in 1951 that no one of them could successfully challenge the power of the companies. In the game the companies played, one country was pitted as a competitor against the others. If one country got out of line, the companies lined up other countries against it and closed ranks to punish the upstart.

The oil-producing countries were beginning to learn that they could improve their position by steadily pressuring the companies and nibbling away at their profits. Once the companies had invested heavily in a country, they were better off continuing to produce even if the host government's benefits increased at their expense. Venezuelan aggressive tax, concession, and royalty policies, the 50-50 profit split, and the deals with independents, playing one company against another, showed that the oil companies were not invulnerable to the assertion of national sovereignty. But no matter what benefits they achieved, the producing countries knew that the companies remained in control of the vital centers of the industry—the price of oil and the amount of production.

Venezuela, the leader in oil policy in the 1940s, realized that cooperation among producing countries could influence the oil companies. If the major producing countries acted together as the oil companies acted together, the divide-and-conquer technique of the companies would not work. Venezuela came to believe that joint action by many oil-producing countries could isolate an oil company, influence it, and through it—the entire industry.

How could the oil-producing countries ever get together? They were competitors. They were scattered all over the globe. Oil had made them rich beyond their wildest dreams, and none wished to jeopardize the flow of money. They had nothing in common except oil. Nonetheless, the idea germinated and grew in the minds of Dr. Juan Pablo Perez Alfonzo and a few others.

These men knew about economic dependence. Their countries, unable to produce manufactured goods, had to import from the same countries that were buying their oil in order to make the goods on which they depended. They also knew that the oil-importing countries had come to depend on oil that they did not possess and that their modern technology required huge and increasing quantities of oil that they had to import. Europe had almost no oil of its own. The United States began importing oil in 1955. Japan had no oil. The fathers of OPEC understood that power was shifting to those countries producing the oil, but that the power was only potential so long as it was diffuse.

As early as 1945, the Arab League regarded Arab oil as one of the Arab nations' sources of influence. Although the league was primarily an expression of concern over Israel, it began to look into oil. Starting in 1952 it had a committee for dealing with oil coopera- tion. Bickering among the Arab nations, however, reduced the benefits of cooperation. In 1947 Venezuela established contact with Iran, hoping to develop cooperation. In 1949 a Venezuelan delega- tion visited Middle Eastern oil exporters, urging exchange of in- formation and cooperation. In 1951 Arab and Iranian officials returned the visit.

For most of the 1950s, the oil business boomed and the producing countries prospered. The exchanges of ideas with Venezuela benefit- ted the Middle Eastern producers through the adoption of the 50-50 formula. In 1953 Iran and Saudi Arabia agreed to cooperate among themselves. Middle Eastern producers began to exchange data and policy moves regularly. Venezuela, however, was in the hands of a dictator friendly to the companies. The conflict with Israel pre- occupied the Arab countries, who, in the 1956 embargo, demon- strated their inability to get together on oil matters. Iran, after the dark night of stagnation following nationalization, was just getting back on its feet. Nothing seemed likely to unite the producing countries.

The producing countries needed only the spark, however, to draw them together. The companies struck the spark in early 1959 when they reduced the posted price. The exporting countries smoldered. They interpreted the move as a raid on their treasuries for the benefit of the companies' profits. They objected almost as much to the manner of the decision as the decision itself, asserting that the companies must consult them before reducing the price. In April 1959, the First Arab Petroleum Congress convened in Cairo. The Arabs invited Iran and Venezuela as observers.

Dr. Perez Alfonzo, once again the oil minister, represented Venezuela. He brought with him his strong feelings about conserva- tion, his hard-line attitude toward the companies, and his country's recent shift from 50-50 to 67-33 in dividing company profits. He had already proposed an international agreement in order to combat the strength of the companies. As spokesman for the world's largest oil exporter, his views carried great weight.

At the congress, Dr. Perez Alfonzo found a kindred soul. Sheikh Abdullah Tariki, the fiery Saudi Arabian oil minister, a graduate of the University of Texas and of Texaco, shared many of the views of the Venezuelan. Although less of a conservationist, he was even more of an ardent anticompany man. They reinforced one another. Before the meeting was over, Venezuela, Saudi Arabia, Kuwait, and Iran had secretly agreed to form a permanent commission to maintain regular contact. Iraq, at the time angry with Egypt, did not attend the congress.

The price cut of 1959 had been only a momentary relief for the companies. Exxon and the others still felt under great pressure. They tried everything to avoid a price cut. In August 1960, however, the oil companies struck flint to stone again. This time the smoldering fire caught on and flames licked the timbers. Action replaced planning. Dr. Perez Alfonzo was lecturing the Texans on conservation, production, and prices. He met Sheikh Tariki at a conference in Midland and the two went back to Caracas together. There on the veranda of the home of Dr. Perez Alfonzo their plan for cooperation jelled.

The August price cut incensed Iraq particularly because it was negotiating with the oil companies and believed that the move was added pressure. Within days of the cut, Iraq sent out the call to Saudi Arabia, Kuwait, Iran, and Venezuela to meet in Baghdad. The mood was right. All the producers were angry and determined to do something. On September 14, the Organization of Petroleum Exporting Countries was born.

THE FOUNDING OF OPEC

In the first meeting of OPEC the oil ministers of the five countries, in a unanimous resolution, resolved:

> that members can no longer remain indifferent to the attitude heretofore adopted by the oil companies in effecting price modifications; that members shall demand that oil companies maintain their prices steady and free from all unnecessary fluctuations; that members shall endeavor by all means available to them, to restore present prices to the levels prevailing before the reduction; that they shall ensure that if any new circumstances arise that in the estimation of the oil companies necessitate price modifications, the said companies shall enter into consultation with the member or members affected in order to fully explain the circumstances. (Cited in Fuad Rouhani, *A History of OPEC*, Praeger, 1971, pp. 77-78.)

Without the diplomatic jargon, that meant no more price cuts, restoration of previous prices, and prior price consultation. These were the immediate goals. Beneath these surface objectives, what the founding members of OPEC really wanted was to increase their revenues from oil. At the moment of the price cuts, the best way to boost revenues seemed to be through restoring the posted price.

Recognizing that all government oil policies could affect their revenues, the second resolution specified: "The principal aim of the Organization shall be the unification of petroleum policies for member countries and the determination of the best means for safeguarding the interest of member countries individually and collectively" (Rouhani, p. 78).

Above all, the OPEC members wanted more money. The years have encrusted more goals and objectives, big and small on the OPEC shield. But central to all the goals and objectives, the strengthening of the economic position of the members looms largest. From the beginning, OPEC has regarded itself primarily as an economic institution whose main mission was to benefit its membership economically. International politics, especially in the early days, played almost no direct role, since the members regarded their adversaries as the oil companies, not other nations or consumers.

The five economically and politically diverse nations that founded OPEC had only crude oil in common. The production heavyweight in 1960 was Venezuela, whose production rivalled that of the next two largest producers—Saudi Arabia and Kuwait. Although it produced 38 percent of OPEC oil, Venezuela's reserve position was the weakest of the five, with only 9 percent of OPEC reserves. The five together exported 83 percent of world exports, produced 41 percent of world production, and owned nearly two-thirds of world oil reserves. Although unable to focus its influence, OPEC, even in its infancy, had the potential force of a monopoly in oil exports.

Since the companies had developed to perfection the technique of dividing and conquering its adversaries one by one, OPEC erected a protective screen. They proposed to defend themselves by unanimity and the agreement not to compete. The first resolution required

> that if as a result of the application of any unanimous decision of this Conference any sanctions are employed, directly or indirectly, by any interested company against one or more of the member countries, no other member shall accept any offer of a beneficial treatment, whether in the form of an increase in exports or in an improvement of prices, that may be made to it by any such company or companies with the intention of discouraging the application of the unanimous decision reached by the Conference. (Rouhani, p. 78.)

The first conference set the condition for the admission of new members. All five founding members must consent. The applicant must export "substantial" quantities of crude oil and have petroleum interests fundamentally similar to those of the original members.

The conference also defined an associate member as a producing country admitted that did not yet export substantial quantities of crude oil.

Three countries became members under these rules. Qatar, always a small crude-oil producer, is a small peninsula sticking out into the Persian Gulf. It became the first admitted member in January 1961. Libya, a new and large producer noted for its valuable light oils, and Indonesia, an old producer but a new, large, and complex country, came in together in June 1962. In 1965 OPEC changed the entry rules. Since that time, a country could gain admission with the consent of three-fourths of the membership, but the consent must always include the original five.

Five more countries have joined since 1962. Abu Dhabi became a member in November 1967. Its membership changed in 1974 when it united with Dubai and Sharjah to become the United Arab Emirates. Algeria joined in July 1969. Nigeria in 1971 became the eleventh member. In 1973 Ecuador and Gabon became first associate members and then full members. They were the last to join. The addition of new members kept the ratio of OPEC to total world exports high—84 percent in 1965 and 1971, and in 1973, 86 percent.

The fuzzy membership criteria—large crude exports and common oil interests—meant that OPEC admitted whom it chose. It systematically excluded developed countries. The Soviet Union put out feelers but OPEC members turned thumbs down. All OPEC members consider themselves poor, less-developed countries whose oil exports—in the hands of foreign oil companies—supported their efforts to promote economic development. Others need not apply.

The OPEC organization itself was spare. Its small office near the Palais de Nations in Geneva housed a few professionals, mostly seconded the OPEC by members. Very early the members displayed their distaste for a large headquarters and a powerful secretariat. The supreme authority and real power of the organization was to be the conference, the semiannual meetings of the oil ministers of members. Members could also call for extraordinary conferences. These meetings, held in exotic places from Bali to Lagos and Quito to Doha, lasted from a few hours to a few days. After the discussions, the ministers voted on proposed resolutions. Those that passed unanimously became the resolution of the conference.

The Board of Governors is the executive authority of the organization. Members elect the board which exercises more immediate control over the organization and supervises the secretariat. The president of OPEC, an honorary office, rotates among the oil ministers. The secretary general, also chosen by the ministers on a rotating

basis among members, serves for two years. He heads six departments—administration, economics, legal, technical, statistical, and
public relations. These departments support the conference with data
and studies, perform housekeeping chores, and undertake general
services. In 1965 OPEC moved to Vienna. Geneva, jaded with world
organizations, refused diplomatic status to the staff, so the secretariat moved to the less sophisticated Austrian capital.

The OPEC organization has no supranational authority. When the
conference passes a resolution, it reflects the unanimous thinking of
the oil ministers but does not bind the members. The conference
transmits each resolution to the members. If none of the members
specifically reject the resolution, OPEC publishes it as the resolution
of the conference. The real power of OPEC therefore lies in the
acceptance and execution of the resolution by members.

The batting average of the oil ministers has been high. Most of the
resolutions have found favor with the members. On some occasions,
however, the members, despite their acceptance, have interpreted the
resolutions to suit themselves, but have usually tried to stay within
the spirit of the resolutions. Members also preserve unanimity by
resolving that each should do what it pleases, as in the royalty dispute with the companies. Members have guarded jealously their
sovereignty and independent decisionmaking. They did not intend
to create an organization that could supercede their national
authority.

BENEFITS OF THE DECADE

The economic benefits to producing countries during the first
ten years of their cooperation were modest but critical. Indeed, without these small economic successes OPEC could not have survived.
The benefits were stable posted prices that were lower than the
1959 level, an improved tax treatment of royalties, and a reduction
of discounts and marketing allowances. All added to the revenues of
producing countries. During the decade the revenues per barrel
increased for OPEC members 17 percent, thus assuring members that
cooperation had an economic payoff, even if modest, and that the
oil companies were listening.

After the 1960 price cut, the oil companies never again reduced
the posted price of crude oil. It would have benefitted the companies greatly to have cut the price further in the 1960s, shifting
more revenues from producing countries to the companies. The
market was in glut during most of the 1960s and the market prices

tended to drift down, reducing total revenues per barrel earned by the companies. As the producing countries received more money per barrel, the companies, in their crude operations, received less.

Some OPEC members made great progress on price actions related to government revenues. Venezuela, with no official posted price, argued that market prices understated the value of its oil. It negotiated with the companies to establish a tax-reference price to replace the market price in calculating taxes and royalties. Beginning in 1966 the companies agreed to pay their obligations using the higher tax-reference prices for five years. But the tax-reference price did not affect the companies' posted price elsewhere. Libya's revenues shifted wholly from market- to posted-price calculations in 1965. In Iran, Kuwait, Saudi Arabia, and other members, the companies had discounted the posted price before calculating taxes and royalties under some circumstances. OPEC pressure helped to eliminate some of these discounts beginning in 1964.

The oil companies at first ignored OPEC and refused to negotiate with it or with individual countries on its behalf. They preferred their old method of one or more companies negotiating with one country at a time. As far as the companies were concerned, OPEC did not exist. As the decade wore on, the companies mellowed, but they never fully accepted OPEC's legitimacy.

The unanimity rule undermined the companies' refusal to negotiate. When a company negotiated with a member of OPEC, it was in fact negotiating with all of OPEC whether it like it or not, according to OPEC doctrine. All OPEC members supported any single member negotiating with the companies. Even though the companies feigned not to see it in the early days, OPEC hovered over the negotiating table.

Agreements between the oil companies and producing countries had traditionally included a reduction from taxable company revenues for the marketing of crude and refined products. The allowance was usually 2 percent of the value at posted price less other discounts. The companies had reduced it to 1 percent even before OPEC came along but stoutly insisted that marketing constituted a real cost.

OPEC contended that the marketing allowances covered costs associated with downstream operations, but not crude-oil production. Costs of marketing should therefore not be deducted from crude revenues. Negotiations began in 1962 and went on into 1963. Finally, OPEC and the companies compromised. The amount deducted by the companies for marketing declined by 70 percent. A small victory for OPEC.

Some other gains came from the royalty negotiations, increasing the revenues of members. Since the beginning of income taxes and the 50-50 profit split, the oil companies had treated royalties as though they were taxes. If the companies were obliged to pay 50 percent of net revenues, they regarded the royalty, say 12.5 percent, as a part of the 50 percent, meaning that income taxes came to 37.5 percent. But if royalties were to be treated as a cost, deducted before determining net revenues, and then the 50 percent rule is applied, the producing countries would receive greater revenues.

Through hard bargaining between 1962 and 1964, the companies and OPEC members compromised on the controversy that was called royalty expensing. The companies agreed to treat royalties as a cost if the producing countries accepted an 8.5 percent discount of the posted price in 1964, 7.5 percent in 1965, 6.5 percent in 1966, with further reductions to be negotiated in later years. With this settlement, OPEC had at last achieved something—but not much—through negotiations.

OPEC ran into some trouble with its rule of unanimity. Some countries were willing to accept the negotiated royalty-expensing formula. Others were not. OPEC resolved the issue by unanimously permitting each country to decide on its own whether or not to accept the compromise. Iran, Qatar, and Saudi Arabia signed. In this issue, it became clear that when sovereignty and independent national decisionmaking conflicted with the majority in OPEC, independence won. The controversy signalled dangers for the future.

The royalty negotiations began again in 1966 and broadened to include discounts for gravity differentials. With the Suez closing and subsequent market improvement, the companies voluntarily improved conditions for the Mediterranean suppliers, especially Saudi Arabia and Libya. The problem remained, however. Discussion dragged on through 1967. Finally, in January 1968, OPEC and the companies reached agreement. The royalty discount would continue to decrease by specified percentages until it reached zero in 1971. The gravity differential would also disappear, but over a longer period. The agreement added to members' incomes. It was for OPEC a clear-cut, if small, economic victory.

OPEC thus achieved some economic benefits for its members in the first decade through negotiations, the pressure of its resolutions, and public relations. Although the companies tried hard to ignore OPEC, they negotiated with it or with countries acting for OPEC. The condition of the market was soft, however, and the bargaining stance of OPEC was weak. OPEC displayed in the 1960s none of the aggressive and determined behavior that marked its negotiations and actions later.

The companies dominated the negotiations and OPEC seemed happy with compromises that conferred few benefits. Rationality and civility marked OPEC's negotiating behavior. OPEC argued from facts and figures and treated the companies with great respect, even awe. The companies played their usual game—delaying, dividing, using their superior knowledge, and employing their economic and political power to achieve advantage. OPEC permitted the companies to drag out the discussion and often succumbed to company power and tactics.

OPEC defended its weak bargaining by arguing that the negotiations were highly technical and it was inexperienced. The members were unwilling to sacrifice their national decisionmaking, to take effective control of crude-oil production, or to install an OPEC production control system. Any of these measures would have amplified OPEC bargaining power greatly. With the market in surplus, market prices declining, and company profits under pressure, OPEC members were still in competition for revenues among themselves to some extent, as well as with the companies.

OPEC performance did improve over the years. By the end of the first decade, the oil companies had begun to accept OPEC as the responsible negotiating authority for the producing countries. OPEC's persistence, moderation, and insistence on negotiating on the basis of facts and figures won the oil companies' grudging acceptance. But much of the modest goodwill that OPEC had earned came from the unwillingness, or inability, to pursue aggressive policies and stir up troubles for the companies.

OPEC PRINCIPLES

Throughout its formative years, the members of OPEC tried to establish and adhere to some strategic principles. Their development was a trial and error process, not codified until late in the decade. The first, of course, was the unanimity rule and ultimate national authority, written into the charter. The second, the most-favored-country principle, sought to give all members the highest benefits achieved by any member from any company. The third, the principle of changed circumstances, argued that prior agreements were subject to renegotiation and change if the positions of the participants and events surrounding the original agreement had changed. The fourth principle was the control of production.

The members have always maintained formal unanimity. Even though it hobbled negotiations on occasion and malfunctioned on the royalty-expensing and other issues, it was the only way the organization could enlist the support of members. If majority or weighted voting had replaced unanimity, some members would have withdrawn. They would not have accepted any formal impairment of their complete freedom of action on oil and oil policy. The fact that OPEC did in fact reduce national authority did not disturb the members. They could and did argue that when this occurred, it was the result of an independent national decision, unforced by an international organization.

OPEC claimed that any benefit or practice conferred by the companies on one OPEC country must apply to all. The companies, of course, had a corresponding principle that all the companies should enjoy the highest benefits conferred by any country on any company. The two principles, direct opposites of one another, vied for ascendancy throughout the 1960s, with the OPEC principle gradually gaining ground. Bit by bit the companies granted nearly uniform treatment to all members.

The oil companies have always been great believers in the sanctity of contracts and agreements. A concession granted in 1908, before oil had been discovered, by a weak or corrupt government must retain its validity for its entire 100-year term. Many of these old agreements were, of course, highly advantageous to the companies, calling for low taxes and royalties and privileged positions. The members of OPEC claimed the right to renegotiate the terms of these agreements on grounds of changing circumstances. Under any conceivable changed circumstances, the producing countries would receive greater benefits.

The two principles—most-favored-country and changing circumstances—interacted. OPEC members noted that Venezuela's more than one-third of OPEC production entitled that country to benefits greater than those received by the two-thirds of production of the other members. This justified, OPEC argued, renegotiations by the other members to bring their benefits into line with those of Venezuela. The companies responded that what Venezuela received from Creole (Exxon) had nothing to do with what Saudi Arabia received from Aramco. Furthermore, the arrangement in each country represented the laws and agreements in each country, not subject to change simply because a country was not satisfied with the deal it had made.

These two OPEC principles played a continuing role in negotiations with the oil companies. Neither side, as a matter of principle, would give an inch. Whenever the companies and countries met to negotiate these propositions were present. Although stoutly defending their positions, the companies were in fact, through the negotiations, accepting the validity of the OPEC claims. The companies, in conferring a benefit such as royalty-expensing, insisted on a quid pro quo, a discount of the posted price. But the discount diminished year by year, so that in the end, the companies had negotiated away a benefit and had treated the members more or less uniformly.

PRODUCTION CONTROLS

Dr. Juan Pablo Perez Alfonzo and Sheikh Abdullah Tariki, responsible more than anyone else for the creation of OPEC, favored production controls and quotas. They knew that the market price reflected supply and demand. Since demand was beyond the control of the producing countries, the only way to insure price stability or increasing prices was to manipulate supply to fit demand at the appropriate price. They assumed, correctly, that during the 1960s, capacity production would exceed demand.

Dr. Perez Alfonzo was especially zealous for production controls since his country—Venezuela—had only small reserves. In 1960 it was already beginning to worry about the exhaustion of its oil. It would serve Venezuela's purpose for all producing countries to hold back production and push the price up—or at least to keep it stable. Under Perez Alfonzo's influence, the first OPEC Conference recommended that "members shall study and formulate a system to insure the stabilization of prices by, among other means, the regulation of production." The second conference in Caracas commissioned a study of international prorationing.

The plight that led to OPEC's birth had resulted from the breakdown of the oil companies' production control system. When the Seven Sisters were firmly in control, they regulated output so that the market remained relatively stable. But during the 1950s, the independents had upset their tranquilizing influence. High production levels and large additions to productive capacity, attributable to the get-rich-quick independents, hung over the market to the degree that the large companies cut posted prices and stimulated the creation of OPEC. The incomplete control of the industry by the major companies transferred the problem of production limitation, at least in part, to the producing countries.

Production-control schemes are extraordinarily difficult to put together and to maintain once established. They are the essence of cartels. The shores of economic history are awash with the debris of cartels, according to economic analysis, because of the failure of production controls. In the usual analysis, the cartels break up because one or more of the participants grow dissatisfied with the assigned share of sales and sell more surreptitiously. The weight of nonauthorized or noncartel production splits the cozy arrangement as competition reemerges. Most cartels thus confer benefits only for a season and then break up. The analogy to the oil situation in the 1950s is incomplete, however, since by that time the cartel had already disappeared, having served its purpose, and the oil companies, each in serving its own interests, served the interests of all, even without a cartel.

OPEC has never created a production-control system. The members in the 1960s examined and studied many proposals, argued about them, passed resolutions, but did nothing. Resolutions at the Ninth Conference in 1965 formulated a specific production plan to rationalize production increases and to diminish oil-company control of production. It even assigned shares of the increase in production to the members. But OPEC members did not implement it.

In the Fourteenth Conference in 1967, Venezuela was back with more pressure for production controls. A resolution reaffirmed the "conviction that a Joint Production Program is an effective instrument for the pursuit of the Organization's fundamental objectives," and decided to study it again. Nothing happened. Venezuela brought it up again in 1969, but still the effort went nowhere. It remains on the agenda and from time to time members discuss it, as they did in Stockholm in 1977. Members, however, are just not willing to give up their national rights.

One obstacle that has prevented OPEC from introducing production controls is how to determine the share for each country. OPEC has considered many criteria, including current and past production, trends, oil reserves, population, income and wealth, financial reserves, and annual bargaining. But always one scheme will benefit one country or group of countries more than another. No country will accept a scheme that does not provide maximum benefits for it.

Other proposals have suggested several criteria simultaneously, giving weights to each formula. Then the weights become the problem. One set of weights will confer greater benefits on a given member than another set. So the members quibble over how much current production or reserves or trends should count in the assignment of shares. With the requirement for unanimity, the jealousy

of national rights, and in a matter of this importance, it is not sur-
prising that OPEC has not reached an agreement.

The real hangup of production controls originates in the constitution of OPEC. No country was willing to sacrifice any part of
its sovereignty or decisionmaking authority to the organization.
A central organization, with real authority, must administer production. Only that would make OPEC a cartel. The shares, once
decided, limit the decisions of each country. Members could no
longer produce what the market and the companies dictated. The
central authority must also police the controls, enforce the shares,
and discipline any member that violates its assigned share. If members had been willing to undertake all this, OPEC would have become a cartel. Venezuela was willing. Other members were not.

The controversy surrounding controls in the 1960s was, in any
case, mainly academic. The oil companies in fact controlled production and exports of each OPEC member. The companies decided
how much to increase production in one field or another in one
country or another, and allocated the oil to their own facilities or
to the markets of oil-importing countries by their own criteria.
OPEC had the potential authority to control production but did not
exercise it.

The company criteria were profits, growth, and stability. Different
companies weighed these elements differently. The oil giants, with
assured profits, weighed stability heavily, holding back the growth
of production in Iran and Saudi Arabia. The independents, struggl-
ing to get a toehold in the industry, placed emphasis on rapid growth
and immediate profits. Libya exploded with new production. The
independents forced the majors to retrench in the interest of stability. But neither the independents nor the majors regarded the
welfare of producing countries as an important consideration in
their production decisions.

The members of OPEC could have established control of production by exercising their national authority, instructing the companies on how much to produce. If the companies did not conform,
the producing countries would then nationalize them, or through
regulation, determine production. In the 1960s not only were the
producing countries not prepared technically to operate their own
industries, but they also feared retaliation and the loss of revenues
if they forced the issue. Iran's experience was still fresh in everyone's
memory. The producing countries felt that they needed the oil
companies. A part of the price was continued company control of
production.

OIL POLICIES

The most important achievement of OPEC in the first decade was the development of oil policies. Before OPEC, each country had its own oil policy. Most were rudimentary, often consisting of little more than the desire for more money through greater production and higher prices. Lacking knowledge of the industry, even in their own countries, oil ministers and other government officials were unable to formulate policies and laws that had a real influence on the industry.

The oil companies were not enthusiastic about helping the producing countries to learn much about the oil business. They did acquiesce in training skilled workers, technicians, and engineering professionals. National personnel, however, seldom entered the management or policy ranks. Whenever possible the companies wrote or influenced the petroleum laws. Ministers usually knew only what the companies chose to tell them. That was insufficient for the countries to establish oil policies.

The companies were not generous in providing production and financial information. What data and analysis they did submit always seemed to favor the companies and cast them in a glowing light. The companies never indicated any need for government oil policies or any intervention in company activities. The vast economic power of the companies provided ample opportunity to influence political leaders and decisionmaking useful to the companies.

One of the principal goals in the early days of OPEC was the unification of petroleum laws through the establishment of a uniform code. Such experts as the producing countries could muster, as well as foreign experts, studied all OPEC members' existing laws and agreements in an attempt to distill their essence. The countries believed that out of uniform laws would come the principles of an oil policy.

The experts concluded that a universal package would be difficult, if not impossible, to formulate. In view of the legal complexity, different legal systems, and unique historical developments in each country, the experts suggested delay and further study. They argued that a body of agreed-upon principles of oil policy must precede and form the basis of uniform laws. OPEC set aside the task until the producing countries learned enough about the oil industry to develop a body of principles.

The process of establishing principles began at once. Resolutions by the dozens poured out of the conferences. Since they were all unanimous, and in most cases the members had ratified them, the resolutions displayed the common ground on which the producing countries stood. Many embodied attitudes toward the companies and their operations, oil resources, legal arrangements, negotiating behavior, and many other topics dealing with oil policy.

In the nineteen conferences before the end of the decade, OPEC adopted 111 resolutions dealing with prices, royalties, taxes, company profits, production volume, support of members' negotiations with companies on many subjects, as well as internal matters. Only two resolutions, toward the end of the decade, dealt specifically with oil policies. But these two organized and codified the policy elements of all the resolutions that had gone before.

Each country made a unique contribution to the establishment of OPEC oil policy. One of the greatest contributors was Venezuela and Dr. Juan Pablo Perez Alfonzo. Venezuela's experience with oil and the oil companies dated back to 1917. It was among the first of the oil-producing countries to become selfconscious about its relations with the companies and the oil market, starting in 1936. It had pushed aggressively on regulation, conservation, prices, taxes, royalties, environmental protection, data reporting, the training of professionals, the reversion to the state of the industry, and many other matters. It came the closest in the 1960s to having a comprehensive oil policy.

The exhuberent Dr. Perez Alfonzo and his compatriots had more experience with oil policy than most public officials. They shared their experience, their policies, and their insights with the others. In OPEC they worked with other countries in modifying Venezuela policies to fit those of the other countries. Sheikh Abdullah Tariki also contributed in the early days. He ceased to be the Saudi Arabian oil minister in 1962 and his place was taken by Sheikh Ahmed Zaki Yamani, one of the giants of OPEC and of OPEC oil policy.

The OPEC Conference and the operations of the governing board and secretariat were one long seminar on the oil industry and how to control it. Producing countries recognized early that they could never hope to control something that they did not know and understand. Conference delegates and their assistants vied with one another in increasing their knowledge of the companies and the industry. The ministries at home, under the impulse of their ministers, who were exposed to international oil problems, became industry workshops. The international secretariat exchanged information, analyses, and views. Officials visited one another's countries, picking up wider experiences, and intimately learned their own industry.

The oil companies often led the seminars. Negotiations with oil men from the companies taught officials of OPEC members the industry's method of operation, its technology, and its problems, as well as how the companies reacted to events and circumstances. Many public officials from OPEC members worked for the oil companies for a time. Nationals of members working for the oil companies often did a stint in the government. Bit by bit, officials from the producing countries penetrated the sacred domain of the companies.

In the course of the decade, OPEC prepared a cadre of professionals who knew the oil business inside out: they were oil men just as much as the company men were oil men. Many were educated in England and the United States and were well-trained economists and technicians. For example, Ali M. Jaidah, secretary general of OPEC from 1977 to 1979, took degrees in economics and petroleum economics at the London School and worked in his home country as a director of the Qatar General Petroleum Corporation. The more that officials from OPEC members learned about oil and the companies, the less they stood in awe of the power of the companies. And the more they learned, the more they knew what they wanted from their country's oil and how they could control the industry.

THE DECLARATORY STATEMENT

The culmination of OPEC's efforts to know the industry and establish an oil policy came in 1968 with the "Declaratory Statement of Petroleum Policy of Member Countries." Sheikh Ahmed Zaki Yamani of Saudi Arabia, who wrote the statement of principles, relied on the entire OPEC experience up to that time. Unnoticed by the press, the statement "was not taken seriously by the oil companies at the time," Dr. Edith Penrose, leading international oil expert, said in her essay in *The Oil Crisis* (R. Vernon, editor; 1976, p. 40). The statement, a unanimous resolution of the Sixteenth Conference in Vienna in June 1968, consisted of ten points, the effects of which, when implemented, would transfer the entire control of world crude-oil production and prices to the producing countries.

The first point the statement made was that "member countries shall endeavor, as far as feasible, to explore for and develop their own hydrocarbon resources directly," relying on private companies only when the government keeps the "greatest measure of participation in and control over all aspects of operations." The second principle goes back to the founding days of OPEC: "the government may require reasonable participation (ownership) on grounds of the principle of changing circumstances."

The third point required the companies to relinquish any concession lands where exploration and production had not yet taken place. Although bound by contractual arrangements, OPEC members also pleaded the principle of changing circumstances in this clause. The fourth principle was a blockbuster. Posted prices "shall be determined by the government," with adjustments for inflation and currency fluctuations. OPEC members not only would take control of production but also, in this provision, of the other commanding height of the industry—the posted price.

The fifth and sixth points established surveillance of company profits. Governments may guarantee the stability of profits only so long as they are not excessively high. The seventh point calls on governments to set accounting standards and require sufficient data from the companies. The eighth point urges governments to determine "the conservation rules to be followed" by the companies. The ninth principle requires company submission to the national courts of producing countries. The tenth point restates the most-favored-country rule by requiring that companies accept the best of current practices in incorporation, labor relations, royalties, and property rights.

The Declaratory Statement was the turning point. OPEC had made up its mind about oil policy. The members wanted everything. The private companies would, as needed, carry out the technical operations of getting the oil out of the ground and shipping it, for fair remuneration. They would not make any of the important decisions, such as how much to produce or how much to charge. Every provision of the statement carried enormous weight, but perhaps the key to OPEC policy was in the first, second, and fourth points; ownership and control of production and control of prices.

To the degree that OPEC members executed their new oil policy, they would create a new and different world crude-oil industry. If OPEC members could, by acting together, force the companies to accept the policy, then the companies' role changed from that of proprietor and policymaker to that of hired technician.

The new policy did not introduce drastic change at once. OPEC members knew that hasty action would meet powerful company resistance. The fruit was not yet ripe for picking. OPEC awaited the changing circumstances that would deliver the industry into its hands. The changes came, sooner than any expected, driven by the desert winds of Libya.

Most of what happened to OPEC in the first decade was known only to a few. The oil companies knew and the producing countries knew. Those who followed the oil industry carefully knew. Secrecy does not explain the lack of knowledge about OPEC. Indifference and the belief that OPEC was a passing mood of oil-exporting countries does. OPEC was a minor momentary irritant that the oil companies would handle. The impact of OPEC was so slight beyond the tight circle of the oil industry that Peter Odell, one of the world's leading oil experts, in his *Oil and World Power*, published in 1970, mentioned OPEC only once in a passing reference.

The 1960s clinched the oil dependence of the industrial countries as the oil industry boomed. The low and stable price, haunted by 15 to 30 percent excess capacity, encouraged the use of oil in industry, transportation, and heating. The extravagant growth of the economies of the industrial countries pushed the demand for oil up and up. The demand for OPEC oil increased 2.8 times, from 7.5 million barrels a day in 1960 to 21.1 million barrels a day in 1970. The aging United States' oil industry increased only 37 percent, to 9.6 million barrels a day in 1970, as that country's imports inched up every year. Quietly and without fanfare, oil became the most important traded commodity, and oil imports became vital to the economies of the United States and Western Europe. How OPEC took advantage of these changing circumstances in the early 1970s is the subject of the next chapter.

Chapter 4

OPEC Gathers Strength

Compared to the calmness, civility, and sedate progress of the 1960s, the first three years of the 1970s were for OPEC a maelstrom of pushing and shoving. OPEC made giant strides toward control of the world oil industry. A radical wing of OPEC developed. Countries such as Libya, Algeria, and Iraq, intolerent of the snail's pace of petty negotiations, wanted to force the companies to yield more of their profits and control immediately. And they did. All the members took a more aggressive stand as the market for crude oil tightened. OPEC members realized that they could control production. They forced the companies to bargain on prices. They brought the industry to the brink of revolution.

Cash and control, the distilled essence of the 1968 Declaratory Statement on oil policy, became the motto of OPEC. When one country achieved a success against the companies, the other countries wanted the same and more. The posted price went from $1.80 a barrel in early 1970 to $3.01 in mid-1973. The "government take" of OPEC members went from less than $1.00 a barrel in early 1970 to $1.80 in mid-1973. Countries bossed their foreign oil companies around and nationalized them if they did not bow. Nationalization, partial ownership, and regulation brought most production under the control of oil-exporting countries. The companies retreated steadily.

Changing circumstances worked their way. Growing oil imports by the United States, Europe, and Japan, inflation in these industrial countries, declining reserves and production in some countries, shortages of crude and products, rising market prices, and the weakening position of the Seven Sisters within the industry provoked OPEC members into a brashness that they had never before displayed. In strong shows of cohesion, OPEC members rallied around each country in the front lines fighting with the companies, even when they didn't sympathize with the country's actions. OPEC proved to the producing countries its worth in solid economic gain.

The power long exercised by the great international companies began to shift to the owners of oil. Three elements entered the equation. First, the large companies, although prospering, began losing decisionmaking authority to the members of OPEC. Second, without suspecting, the independent companies had weakened and continued to sap the strength of the large companies, making it impossible for them to retain control by acting together. The third element was the rising market. Surging demand moved through the intermediaries, large and small, to the ultimate suppliers, and added to their growing strength.

LIBYA DIVIDES AND CONQUERS

Libya and the turn of the market radically changed the circumstances. That poverty-stricken North African Arab country did what the older and wiser oil-producing countries failed to do. It threw caution to the winds and challenged the oil companies directly, threatened and castigated them, and got its way. But all through the turmoil of 1970 and 1971, other OPEC members lent their strong support. The once-in-a-lifetime set of circumstances, as well as the lack of foresight on the part of the oil companies also helped the OPEC cause. Still, it was Libya's initiatives that in 1970 set in motion the events that culminated in the OPEC price revolution in the fall of 1973.

Libya was a newcomer to oil. It had astutely opened the country to competitive bidding for concessions. It did not wish to depend on only one company or a small group of companies and was anxious to get production started and the money rolling in. The independents as well as the Seven Sisters flocked to Libya in the 1950s. But Libya nullified its auspicious start by bargaining poorly on the concessions and by failing to watch the companies once they got in. Libya permitted the market price, rather than the posted price, to determine its revenues for the independents. It also allowed the companies to rush headlong into large-scale production without adequate conservation safeguards.

The oil giants as well as OPEC supported Libya when in 1965 the new petroleum law changed the rules to base revenues paid by independents on the posted price. The large companies regarded the independents as obnoxious interlopers. The independents viewed the large companies as cartel members determined to squelch them. This antagonism between the independents and the majors later undermined the one chance that all the companies had to avoid their mutual disaster. In 1966 when Libya opened an even greater concession area, more companies, big and small, swarmed into the country. In just a few years, Libya became an important contributor to the European oil market.

Libya's independents added to the woes of the large companies. The new independent production upset the delicate balancing act of maintaining stable growth acceptable to all the giants. The independents extracted the oil from the ground as fast as they could, often ignoring the proper care of the oil fields. In their eagerness, they sometimes even forgot to meter the oil as they loaded it for export. The independents showed but small concern either for the market of for Libya.

Libya was boom town that contained great hurrying, scurrying, and shortcuts by the companies. Many of the companies were new and inexperienced, and all were thirsty for quick profits. The companies did not feel threatened by the Libyan political regime, which was most anxious to receive the oil money. The government, under King Idris, was benign and friendly, although it was pressing for a modest increase in the posted price.

All that changed suddenly on September 1, 1969, when Colonel Maummar al-Qadaffi and his coterie of young army officers deposed King Idris and set up a puritanical government. Full of Arab nationalism and hatred of Israel, the colonel and his men lost no time in letting the oil companies know that Libyan oil would cost them more. The civility, rationality, and casual attitude of the King's government vanished. Colonel Qadaffi would rule or ruin.

The new government called the twenty-one foreign companies on the carpet and demanded a higher posted price and more taxes. Colonel Qadaffi at first zeroed in on two companies—Exxon and Occidental—demanding 40 cents more per barrel on the posted price and 55 percent instead of 50 percent in taxes. The companies had difficulty adjusting quickly. They were still thinking of the 10 cents a barrel that King Idris had wanted. They just didn't take Colonel Qadaffi seriously at first.

Libya and its neighbor, Algeria, interacted, reinforcing one another's radicalism and demands for higher prices and taxes. Algeria was giving the French, who held 70 percent of the country's production and reserves, a hard time. It was also demanding a much higher posted price from Royal Dutch-Shell and Phillips. When they delayed in 1970, Algeria nationalized them and began making noises about nationalizing the French companies. J. Paul Getty, when he received a new concession in 1968, gave Algeria more than had the other companies and Algeria tried to instigate competition among the companies. Algeria's militant stand toward the companies encouraged the new government of Libya. Later, Algeria followed Libya's example by forcing its companies to pay more.

In Libya the companies stalled, as they always had, hoping that the negotiations would drag on and on, which would be to their advantage. But Colonel Qadaffi wouldn't wait. He moved quickly against Occidental—also a newcomer in oil. Occidental was an independent and operated crude-production facilities only in Libya for its European refineries. Its freewheeling owner, Dr. Armand Hammer, had bought the moribund California company and was making a fortune supplying the prosperous European market. He had no intention of sharing his profits with Libya.

In May and June 1970, Colonel Qadaffi ordered Occidental to cut its production by almost one-third. He said he ordered the reduction for conservation purposes. Occidental, like many other companies, did in fact engage in questionable conservation practices. But the real reason was to add pressure to the country's demands for more taxes and a higher posted price. When Occidental dallied in complying with the cutback, Colonel Qadaffi threatened to send his troops into the oil fields. Dr. Hammer complied.

Desperate for crude oil to feed his booming European refineries. Dr. Hammer appealed to Exxon and the other giants for enough crude oil to tide him over the shutdown. Exxon and the others, however, were miffed at Occidental in particular and the independents in general for their troublemaking and irresponsible increases in production. Exxon blandly offered to supply Occidental with crude at the going third-party rate. That was a slap in the face for Dr. Hammer. Exxon just could not see that what was happening to Occidental would also happen to the other independents and finally, to Exxon. Negotiating from weakness, Dr. Hammer capitulated in August, agreeing to pay 30 cents more for crude and 58 percent of net revenue in taxes.

With the Occidental victory under his belt, Colonel Qadaffi then attacked the Oasis group, a consortium of three independents and Royal Dutch-Shell. The independents caved in quickly, accepting the Occidental terms. But Sir David Barran, chairman of Royal Dutch-Shell, was more astute than Exxon. As he related in a letter to Senator Church (included in the *Multinational Hearings*), he believed that his company should "at least try to stem the avalanche," and he refused to give in. It was a quixotic gesture.

In late September Colonel Qadaffi halted production in the Royal Dutch-Shell part of the Oasis concession. By this time, only the large companies—Exxon, Royal Dutch-Shell, Mobil, British Petroleum, Texaco, and Socal—were holding out against the colonel. The independents were back in production, on Colonel Qadaffi's terms, having received no help from the large companies. The choice of the majors now was to close down in Libya or to accept the colonel's price.

If the large companies defied Colonel Qadaffi, he would either nationalize them or shut them down. The independents and the national oil company that took over the majors could then create havoc in the European market. If they gave in, other producing countries would pressure them for equivalent benefits. Their traditional defenses—boycott and isolation—would not work here as they had so effectively in Mexico and Iran. Indeed, Libya was turning their own divide-and-conquer methods against the companies.

Matters of principle that cost money do not stand high with oil companies. First Texaco and Socal buckled, then the other American sisters, and finally Royal Dutch-Shell. Colonel Qadaffi had routed the companies and stirred up a hornet's next, both in OPEC and in the oil market. The large companies feared new demands from other countries—the leapfrog effect. They waited uneasily for the Persian Gulf OPEC members to drop the other shoe.

Despite his militancy, Colonel Qadaffi negotiated with the oil companies. Indeed, he imposed his will unilaterally on the companies only when they refused to negotiate or violated his orders. And his orders were always scrupulously correct according to Libyan law. The intense discussions between the companies and the government demonstrated abundantly to the companies that circumstances had changed and that Libya intended to reap the rewards.

LIBYA'S VICTORY

Colonel Qadaffi's boldness and moral courage explain only a part of his victory. The changing circumstances and particular events had rallied to his aid. The presence of independents that required Libyan crude oil undermined the position of all the oil companies. The absence of crude oil from any other source to satisfy the prospering European market made the Libyan supply crucial.

The Suez Canal had closed in 1967, shutting off most Persian Gulf crude. The other method for getting Persian Gulf crude to Europe, the Trans-Arabian Pipeline (Tapline) that delivered 500,000 barrels of Saudi Arabian crude daily to Europe, suffered a bulldozer accident in May 1970. That was the exact moment that Colonel Qadaffi was closing down part of Occidental's production. The Syrians would permit repairs only after negotiating higher payments from the companies controlling the pipeline. That took time. Meanwhile, Tapline was closed down.

The closing of the Canal also led to a shortage of tankers to bring in outside crude. The Biafran rebellion had forced the suspension of Nigerian crude production in the summer of 1970. Finally, Colonel Qadaffi counted on the shortsightedness of the large companies. He was right. They did not establish common cause with the independents to frustrate his divide-and-conquer attacks on the weak links in Libyan production.

More than just a spot European shortage of crude oil favored the Libyan gambit. The whole oil market was wheeling around to larger deficits. The galloping growth of consumption in all the oil-importing countries had finally caught up with production and had sopped up excess capacity by 1970. From 1965 to 1970 world imports had increased 71 percent and OPEC exports 69 percent. Libya's exports had nearly tripled. Years of car building and use, of industrial technology using oil-intensive equipment, and of heating and electrifying the world began to press on supplies. Since Europe had rebuilt its war-torn industry with oil-burning equipment and was emulating the American mania for cars and trucks, it was the first to feel the pinch.

The year 1970 was the peak production year for the United States, still the largest producer as well as consumer. Its imports of oil and the import share of its consumption were climbing each year. The United States needed all of its own oil and then some. Many oil experts thought it was time to raise the posted price. The foreign offices of oil-importing countries did little to support its oil companies, partly to assuage Libya and partly because they thought Libya was right. Oil men everywhere felt tightness in the market, and prices were moving up. But the companies were reluctant to share the improved conditions with the oil-producing countries.

Colonel Qadaffi's success quickened the pace in OPEC. Soon,

other OPEC members, especially in the Middle East, were demanding a higher posted price and more in taxes, the same as or even more than Libya got. Each member wanted at least the benefits going to the most-favored nation. The Persian Gulf countries were especially insistent. Their caution and conservatism had earned them some respect from the companies, but not much money. They decided to get tough. The avalanche picked up speed.

Excitement and change vibrated in the air when OPEC members assembled in Caracas in December 1970, for their Twenty-first Conference. Just a few days before the conference, the Venezuelan Congress had passed a law giving President Rafael Caldera the sole authority for setting the tax-reference price, the Venezuelan version of the posted price. Since 1966 the companies and the government had negotiated the tax-reference price. OPEC technical studies presented at the conference showed that the market for crude oil was improving rapidly, that the long glut had passed, and that a long period of shortage was coming.

The conference resolved that circumstances had changed and demanded 55 percent as the minimum income tax, as well as a higher posted price, for all members. It requested early negotiations in Teheran with the companies. The oil ministers made it plain that the posted price must go up. The unanimous resolution spoke of "simultaneous and concerted action," code words, as Edith Penrose says in her essay in *The Oil Crisis*, for a worldwide crude shutdown that the ministers discussed openly. The OPEC worm had turned.

PREPARATIONS FOR TEHERAN

If the Libyan affair was the first stiff wind whistling across the desert, then Teheran and Tripoli in early 1971 were the sandstorm in full force. Agreements in these cities shredded company strength and unity and placed the OPEC forces in joint command with the companies of the world crude-oil industry. The leapfrog that the companies so greatly feared not only jumped the price, but also plunked OPEC down alongside the companies at the price bargaining table.

By a combination of intransigence, threats, unity, and astute bargaining OPEC members prevailed against the oil companies. Self-interest divided the companies. Neither their governments nor a buyer's market supported them. In the negotiations, they stood on shifting sand. When the storm subsided in April 1971, the posted price had advanced more than one-sixth and OPEC members' government take had increased by more than one-fourth.

The ominous sounds from the Caracas Conference produced rumors that producing countries were really considering a world crude shutdown. In 1971 the oil companies knew that this was a possibility, so vivified and unified was OPEC following the successful Libyan belligerence. In the prosperous oil market the importance of the Teheran meeting spurred the companies to elaborate preparations.

The companies feared three events. They were alarmed by the possible effects of higher crude prices and taxes that would come from another OPEC victory. They knew that they could pass the increases along, but that took time, reduced profits, and would arouse popular sentiment. Their home governments and customers would accuse them of playing footsie with OPEC.

Their second fear was of more leapfrogging. Libya had set the stage by obtaining a higher price and more taxes on Mediterranean crude. Now the Persian Gulf countries wanted more. If they got more, or even the same, the Mediterranean producers would be back again for more. Then the Gulf would want more, and so on and so on. The instability of the industry would damage the companies. Their third fear was the disorder in the market that would follow in the wake of the negotiations. Price increases and higher taxes would create market instability, provide yet more opportunities for the independents, and the scramble would threaten market shares, planning, and orderly growth.

The Seven Sisters and sixteen other large oil companies met in New York to set their snares for OPEC. They resolved first to join forces and to bargain only collectively with all of OPEC at the same time. There is a certain irony in this. Only a decade earlier, the companies insisted on talking to OPEC members one at a time. Now they felt that they must bargain with all simultaneously. Circumstances indeed had changed. By talking with all of OPEC at once, they believed that they would fortify their negotiating strength and leave no room for divisive, parallel, or successive negotiations, scotching the Persian Gulf-Mediterranean leapfrog.

Twenty-three companies penned a joint letter to OPEC saying that they could not further negotiate the development of claims by OPEC member countries on any basis other than one that reached a settlement simultaneously with all member countries. Then the companies rallied the support of the State Department and the foreign offices of other importing countries to strengthen their case politically.

To give force to their position, the companies all agreed to come to the aid of any distressed company singled out by an OPEC member or members. This was a safety net for all companies. In effect, the companies agreed to pool their crude. The antagonism between the giants and the independents did not vanish, but both saw that only through cooperation could they hope to prevail. The failure to develop a safety net had contributed greatly to the debacle in Libya. The agreement was secret, not revealed until 1974, but OPEC presumed its existence.

The companies secured a letter permitting joint negotiations from the Anti-Trust Division of the Department of Justice. They also created the London Policy Group, as well as an informal group in New York, consisting of phalanxes of the major companies' top executives who would parry the OPEC thrusts and direct the battle. These groups were the second line of defense and were in direct touch with home government officials. The companies now thought that they were all set to beard the sheikhs and the shah.

OPEC members hit the ceiling when they received the joint letter from the twenty-three companies. Although they had wanted to bargain collectively with the companies ten years before, now their advantage had shifted to separate negotiations. Incensed, the Shah of Iran and his finance minister, Dr. Jamshid Amouzegar, a Washington University hydraulics engineer, insisted on separate Persian Gulf negotiations. Then Libya refused to participate in the Teheran meeting. OPEC appointed Sheikh Yamani, of Harvard and New York University, Saadoum Hammadi, chairman of Iraq National Oil Company and a PhD in agricultural economics from Wisconsin, and Dr. Amouzegar as negotiators. They announced that collective negotiations were not acceptable.

The company plan for collective negotiations unravelled. The American government weakened after its ambassador had a short unpleasant talk with the Shah of Iran. Then the London Policy Group equivocated. Finally, amid disarray and confusion as to exactly what was happening, the London Policy Group, under threat of a shutdown, and with only two days to decide, acceded to "separate (but necessarily connected) discussions," as shown in *Multinational Hearings* (part 5, p. 132). That really meant that the Gulf and Mediterranean OPEC members would negotiate separately and in sequence. OPEC had won.

THE TEHERAN SHOWDOWN AND TRIPOLI

The Teheran meeting got down to business on January 28, 1971. George Piercy of Exxon, new to the job, but an experienced oil man, and Lord Strathalmond, managing director of British Petroleum, lawyer and son of the previous director of British Petroleum, negotiated for the companies. They allowed five days for a settlement. The companies began by offering 15 cents more a barrel and a modest annual increase for inflation.

Dr. Amouzegar countered for OPEC with 54 cents more per barrel and a much higher inflation allowance. The producing countries again threatened a Persian Gulf crude-oil shutdown. The companies had no threats to make. Both sides were adamant. But the pall of defeat hovered over the company side of the table. They could not afford a shutdown. After five days the negotiators were still far apart and the companies adjourned the discussions.

OPEC held a conference on the spot. They resolved that if by February 15 the companies did not accept a negotiated settlement, each member would legislate new prices and taxes for its own oil. And if the companies didn't accept the legislated terms in each country, OPEC would close down the whole industry. OPEC had the companies over a barrel.

On Valentine's Day, the companies, tails between their legs, came back to the table, negotiated, and signed. The posted price for Persian crude would go up 30 cents a barrel immediately and would advance 20 cents more over the succeeding five years for inflation. The profit split would change for all members to about 55 percent for governments, 45 percent for the companies. The decisive victory for OPEC demonstrated without doubt that the balance of power in the industry had shifted from the companies to the producing countries. Equally important, the companies could no longer fix the price without negotiating with OPEC.

The Mediterranean OPEC members now prepared to leapfrog over the Persian Gulf states. Saudi Arabia and Iran, participants in the Persian Gulf negotiations, joined Libya and Algeria of the Mediterranean group, since Tapline gave them access to the inland sea. Nigeria, another Mediterranean supplier, picked this moment to join OPEC, strengthening the OPEC side even more. Libya's Major Abdul Salaam Jalloud negotiated for OPEC. At first, he refused to talk to the companies as a group, although George Piercy was on hand for that purpose. Major Jalloud called the companies in, one at a time, and scolded and humiliated them, demanding more than the Gulf countries had obtained. It was OPEC's turn to show contempt for the companies.

Major Jalloud flung one company offer to the floor and watched the company spokesman stoop to pick it up. Another company man, unable to get an appointment to see the major, slipped his company's offer under the door of the oil ministry. Finally, in March, Major Jalloud accepted joint discussions. By this time, many of the companies, especially the big ones, were ready to give up. Their principal stake was not in the Mediterranean and they wanted an agreement—almost at any cost. On April 2, 1971, the negotiators reached a five-year agreement that was better for OPEC Mediterranean members than the agreement made at the Teheran meeting. Countries not in the Gulf and Mediterranean groups came to terms with the companies on similar terms. The leapfrogging stopped.

Neither agreement took into account the event that transpired in December 1971. The United States reduced the value of its dollar relative to other currencies by about 9 percent. At the same time, European currencies increased in value not only by the dollar de-valuation, but also by another amount decreed by several govern-ments. Since the oil-importing countries paid in dollars, the currency changes substantially reduced the purchasing power of the dollars that OPEC members earned. Members now wanted more dollars to offset the currency changes.

OPEC called the companies to a meeting in January 1972, to correct the oversight in the Teheran and Tripoli agreements. The companies didn't even put up a fight. They agreed to link the posted price to a group of currencies, not the dollar alone, in effect raising the price in dollars. The position of the dollar continued to weaken. The currency basket protected OPEC to some degree. But the United States again reduced the value of the dollar in February 1973, as European currencies once again increased in value. The companies came back to Geneva in June 1973, to increase the dollar value of the posted price even more than before and to arrange for OPEC a more favorable basket of currencies.

Between 1970 and 1973, when OPEC members had the oil com-panies on the run, the oil market tightened even more and market prices went up. Worldwide inflation, a problem since 1965, went rampant, fueling OPEC demands for higher inflation allowances. Even with the negotiated posted price increases, oil products, com-pared to most other products, were still a bargain. The higher oil-product prices had not discouraged demand. Indeed, consumption of oil products seemed insensitive to price increases, but were indeed sensitive to the growing incomes of the industrial countries.

Exports of OPEC members went up and up. Although American, Libyan, and Venezuelan production had peaked in 1970 and had gone down thereafter, other producers took up the slack. Saudi Arabia emerged as the largest producer and exporter. OPEC exports increased 12.3 percent in 1971, 7.3 percent in 1972, and 16.0 percent in 1973. Saudi Arabian exports doubled. Neither the higher price nor the warning signals of shortages in the industrial countries served to restrain the world's appetite for oil.

Company after company restudied their supplies, reserves, and markets, about which they had glibly reported surplusses earlier. Now they found not temporary, but long-term shortages. The bottom of the barrel was in sight. Government studies also revealed increasing tightness in the oil market. Experts began to compare reserves to consumption rates. Soon they spoke of running out of oil by the end of the century and constantly rising prices. Governments of oil-importing countries began to formulate oil and energy policies designed to encourage production, discourage consumption, and seek out substitutes.

NATIONALIZATION

The second great application of the principle of changing circumstances conferred the ownership of crude oil and production facilities, the upstream end of the oil industry, on OPEC members. Concessions granted by producing countries in bygone times effectively vested ownership of the oil in the companies. Only twice had the producing countries challenged that arrangement. Mexico and Iran had taken ownership by nationalizing their companies. But the company buzz saw had cut them down, forcing one out of the export business and the other into a partnership that was little more advantageous than the previous concessions.

With growing influence over price, OPEC members now decided to chance more nationalizations. Some were coldly calculated, undertaken in order to achieve control of the nation's oil industry. Others resulted from company misbehavior in the view of the government or company disobedience of government orders. Still others were motivated by political considerations or a pique of the moment.

Equally important, perhaps even more important for the private industry, some OPEC members pressed for the forced but negotiated purchase of producing oil companies by OPEC members. OPEC called this procedure participation. The two methods together represented a major step toward the fulfillment of the 1968 policy statement that included increasing control by the producing countries of their own oil reserves and production.

Some countries preferred the nationalization route. Libya nationalized British Petroleum in a fit of anger over the Iranian occupation in 1971 of some Persian Gulf islands under British protection. A year and a half later, it took over 51 percent of Bunker Hunt interests. Just three months after that, in September 1973, Libya claimed 51 percent of the holdings of Exxon, Royal Dutch-Shell, Mobil, and Texaco.

Venezuela did not nationalize its companies until 1976, but in 1972 it laid the groundwork. A new law insured the reversion of all private facilities to the government in 1983 without compensation. The companies—Exxon, Royal Dutch-Shell, Mobil, Texaco, and ten others—had to post bond that guaranteed that the industry would not deteriorate in the interim. No sooner had the President of Venezuela signed the reversion law than the government started making plans for even earlier nationalization.

In February 1971, Algeria, after toying with French holdings for more than five years, finally nationalized them, ending that country's hopes for secure crude supplies. After a decade of quarreling that included government recapture of unworked concessions, Iraq in June 1972, fully nationalized Iraq Petroleum. In these nationalizations, the oil companies could not mount effective retaliatory boycotts or demands for concessions beyond those offered by the countries. In most cases, producing countries paid or arranged to pay the net book value, much below the worth of the properties.

Even though some of the nationalizations were far from amicable, many of the companies remained in place. In some cases they stayed as minority owners of the facilities, having lost only the control of their enterprises. In other cases, the companies established service companies that bought the crude oil from national companies that they had formerly owned. The companies continued to have access to the crude they needed. They did not always have preferred access they would have liked.

Saudi Arabia and other Arab countries took the participation route. Back in 1921 Great Britain and France had guaranteed Iraq 20 percent of Iraq Petroleum while the company was still in process of formation. Later, they reneged on the agreement and left the Iraqis out. That breach of faith had been a sore point between Iraq and its consortium over the years. The pledge to permit participation, designed to enlist the country's support, pointed the way, decades later, for other Arab countries, even though Iraq had chosen to nationalize.

By the 1970s Sheikh Yamani of Saudi Arabia had picked up the idea of participation. Although he had been named a director of Aramco in 1962, the company in fact excluded him from any real authority. The ink was barely dry on the Teheran Agreement when he began to press for an immediate 20 percent Saudi Arabian share of Aramco, a percentage that would gradually rise to 51 percent in a few years. Not surprisingly, Aramco resisted.

Participation exposed the companies to risks not inherent in nationalization. With nationalization, the oil company was out. It could take its payment and go elsewhere. If the nationalization was only partial, the company could sell out or take its licking and just pull out. The company could, of course, also choose to stay and buy the oil it once owned, although the nationalized companies could choose to sell their oil elsewhere.

With participation the company and the country became bedfellows. Each shared the same interests. Increased benefits to the oil companies meant increased benefits to the producing countries. When the oil came out of the ground, a specified percentage belonged to the government. The company sold it, along with its own oil, to itself or to the downstream companies and paid the government the negotiated price. The company and the countries in the participation arrangement were wed, a feature of participation that had not escaped Sheikh Yamani, who told a reporter that participation "would be indissoluble, like a Catholic marriage" (cited in Anthony Sampson, *The Seven Sisters*, p. 276).

The events of the early 1970s were beginning to change the minds of the oil companies. They had learned to accept OPEC participation in the price setting and the higher price of crude oil, reasoning that they could pass the higher price through to retail markets. The larger government take would force them to shift their profit center from the crude-oil operation to the refining and distribution operations. The higher crude price would cut sales slightly, and the market adjustments did not threaten their survival. The companies began to realize that the ownership of the oil was not the critical element. In the old world of supine producing countries, ownership was best. It gave the companies the crude and kept other companies from getting it. But in the new world of pushy countries, the most important considerations, especially to the large companies, was the ability to continue to buy the oil.

Joint ownership with the oil-producing country guaranteed their crude supplies, since the country offered preferred access. The country might take a modest amount to market separately, but the success of the private company also spelled the success of the country. Nationalization did not have this built-in guaranty and indeed did threaten their survival as oil companies. So the companies accepted that participation was second-best to ownership and was often better than nationalization.

PARTICIPATION

The discussions between Sheikh Yamani, the spokesman for the Arab Gulf countries on participation, and the companies dealt first with Aramco and began in 1972. Whatever deal Aramco and Saudi Arabia made, the other companies and countries would accept. First, Aramco suggested a 50-percent share for Saudi Arabia in future oil development and dangled a multi-billion dollar investment program before the country. Sheikh Yamani scoffed at it.

Aramco inspired an appeal by President Nixon to King Faisal, protesting the stiff terms and asking the king to use moderation. Sheikh Yamani, annoyed that the company tried to go over his head, appealed to his king and brought back a threat from King Faisal: unless Aramco accepted Saudi Arabian participation soon, his government would take unilateral action. That meant the nationalization that Aramco had come to fear.

Still the discussions dragged on through 1972 with threats and counterthreats and obdurate positions taken and forsaken. Sheikh Yamani held the strong hand. Both he and Aramco knew it. Aramco desperately needed Saudi Arabian oil and was just as anxious to keep it out of the hands of other companies. Saudi Arabian oil was by now a significant part of the business of the individual companies that owned Aramco—Exxon, Texaco, Socal, and Mobil.

Soon the battle was over, and only the terms of Aramco's surrender remained. In October 1972, Aramco's owners signed away control of the company. Saudi Arabia would receive immediately 25 percent of the concessions and production facilities, and therefore that share of current oil production. The percentage would rise year by year to 51 percent in 1983. Soon Kuwait, Abu Dhabi, and other producing countries extracted the same terms from their companies. The camel entered the tent, head and shoulders, on the first try.

Participation did not settle the whole matter. Saudi Arabia, of course, did not want its one-fourth to one-half of Aramco production. The country wanted Aramco to have it—but at a price. Sheikh Yamani and the companies went back to the bargaining table. Some of the companies wanted to arch their backs at this point and insist on a heavy discount of the posted price for the government oil. But the same urge to have a guaranteed source of crude oil that impelled them to concede on participation itself now forced them to accept Sheikh Yamani's buy-back price. But not without at lot of talk.

Sheikh Yamani watched and waited. At one point, patience

thoroughly tried, he threatened to order Aramco cutbacks. With the crude market booming, Aramco was horrified. Finally, in mid-September 1973, Aramco accepted Sheikh Yamani's buy-back price, 93 percent of the posted price. The camel was now munching comfortably on all the stores in the tent.

The United States and other oil-importing countries were enjoying high prosperity and rapid economic growth. Governments were aware, of course, that the oil market had turned and that some oil exporters were flexing their muscles. The State Department, however, enjoyed good relations with Saudi Arabia, Iran, and nearly all the other members of OPEC. And the government had great confidence in the ability of the oil companies to handle any recalcitrant oil-exporting countries. The U.S. government was preoccupied with national security, the war in Viet Nam, and the unravelling Watergate scandal. Minor summer gasoline and winter natural gas shortages were annoying, but few believed that an oil crisis was in the making.

By fall of 1973 the stage was set, however, for the denouement in the oil industry. Both the oil companies and OPEC members knew that change was in the offing. But neither knew exactly what lay ahead. The oil companies were demoralized, defensive, and fearful. OPEC members were riding high, confident that history for the first time was running in their direction. Most of the world was unaware of what had already happened or of the positions of the power contenders. Oil consumers realized vaguely that oil was less plentiful and higher priced than before, but were blissfully ignorant of the struggle going on in the industry.

The tiny flame that ignited when five countries that counted for nothing in the world power balance formed OPEC in 1960 had become a blazing inferno. It needed only a gust of wind to engulf the whole world oil industry. The breezes whipped up by the churning market pushing the market price above the posted price, along with the backwash of rumbling tanks in the Sinai, was enough.

The revolution, when it came in October 1973, was quick and decisive. It lasted only through October, November, and December. An earlier chapter has told the story of those three months. But successful revolutionaries do not always hold their ground. The next chapter shows how OPEC has coped with its success, how it has managed the industry, and how it has used its power in the years since 1973.

Chapter 5

Since the Revolution

"On October 16 we got as great a shock as you did. We thought we were pygmies facing giants. Suddenly we found that the giants were ordinary human beings; that the Rock of Gibraltar was really paper mache," Abdullahif al Hamaad, Director of the Kuwait Fund, told Anthony Sampson. The English journalist relates the conversation in his *Seven Sisters* (Viking Press, 1975), one of the best books in recent years on the oil industry. OPEC had marched across contested ground in the last three months of 1973, trampling underfoot both the oil companies and the oil-importing countries. By January 1974, it occupied the territory. But would OPEC hold the ground? Would counterattacks by its adversaries break OPEC? Would OPEC mismanage the industry and lose its grip? Would the condition of the market push the oil-exporting countries once again into a subordinate position?

The most dangerous moment for most revolutions is often the time immediately following its success. In the OPEC case, discord ripped into the unity that won the victory as the leaders jockeyed for place and power. The reappearance of normality and stability brought back all the old jealousies and national ambitions. Controversies over the price and shares of exports that OPEC had papered over with the heightened unity of the drive to power erupted once again. Inexperience and uncertainty at the controls added to OPEC's woes, as did the fear of retaliation.

OPEC held its ground. The oil companies made their peace quickly, learned to cooperate, and even helped OPEC in getting the oil to market at the OPEC price. After some threats and blusters, the United States and other oil-importing countries accepted OPEC's price setting and control of production. OPEC mastered enough of the economics, finance, politics, management, and technology of the crude-oil industry to operate it effectively in its own interests. Since 1973 OPEC members have continued to prove that producer cooperation can be immensely beneficial. Despite dozens of ideas to bring OPEC down, its control has not faced a serious challenge.

Since 1973 the market has gone up and down with only a slight upward trend. The occasional sluggish market tested the determination of OPEC members to permit their exports and government revenues to vary while holding the price at the level they established. Excess capacity, reaching nearly one-third of exports in 1975 and prevailing to some degree most of the time, tempted members to cheat. Some indulged in discounting in order to maintain their exports. The real or relative price—the price of crude oil relative to other goods—has declined since 1974. At one time the members split on what price to establish, but the breach was quickly healed. OPEC worries about the deterioration of the dollar and inflation at the same time that the oil market and excess capacity permit modest price increases. Even so, the lure of money, the absence of effective external threats, and the taste of power keeps OPEC strong and in control. The return of the strong market in 1979 brightens OPEC's prospects.

EFFECT ON WORLD TRADE

The immediate effect of the oil-price revolution on world commerce was economic and financial chaos. World finance had to transfer huge amounts of money to OPEC members. In 1974 earnings tripled. The world financial community had never before received such a jolt. Surpluses became deficits and slight deficits of some importers became impossible deficits. Countries borrowed frantically to pay their oil bill. Some worried that the financial system might collapse. It didn't. Everybody rallied around. Finance ministries of all countries, banks, and international institutions saved the day. OPEC treasuries shoveled their money back into the system by increasing imports and investing in world capital markets.

World economic growth skidded to a halt. The countries of the Organization for Economic Cooperation and Development (OECD)— the industrial countries—grew in 1973 at 6.2 percent, but had a growth rate in 1974 of 0.4 percent. The United States and Japan lost ground, minus 1.4 and 1.1 percent respectively. The less-developed countries continued to grow, thanks to some raw-material price increases. But their trade deficit went from $12 billion in 1973 to $33 billion in 1974. Noncommunist world product increased only 2.2 percent, compared to 6.7 percent in 1974. OPEC national product, on the other hand, increased 250 percent.

The next year, 1975, was even worse. The gross national products of OECD and less-developed countries lost ground, the latter by 7.7 percent. Noncommunist world product declined 1.0 percent. The two largest less-developed countries, Brazil and India, marked time. Even OPEC members stood still, for that was the year their exports went down and earnings fell because of the world recession. The trade deficit of the less-developed countries hit an all-time high of $44 billion.

For world trade and development, 1974 and 1975 were two very bad years. The developed countries were hit hard and the less-developed countries were doubly brutalized. The recession in the industrial countries dried up their markets, and OPEC markets were not promising. At the same time, the price of oil drained foreign exchange from less-developed countries, raised costs, and made borrowing mandatory. By 1974 the less-developed countries owed, in foreign public loans, $113 billion. The next year, the debt was $144 billion, with a debt service of $21 billion—one-fourth of their exports. Raw-material price rises in 1973-1975 saved some less-developed countries, but others faced disaster.

Since 1975 the world's economies have improved. Growth has resumed, but at a much slower pace. Industrial countries grew at 5.0 percent in 1976 and 4.0 percent in 1977. In 1973-1978 the growth of the major industrial countries was less than one-half of what it had been in the five years up to 1973. The less-developed countries struggled back to their 1974 production in 1976 and in 1977 and 1978 made some progress. Their trade deficit declined to $27 billion 1977, but external debt continued to grow, reaching more than $240 billion in 1979. World inflationary pressure eased somewhat, but not by much. The world, by 1979, had absorbed most of the economic shock of the oil-price revolution of 1973-1974 and had resumed its old ways.

Some things can never go back to the way they were. Oil and everything made with oil are now more expensive. OPEC members, because of their market power, in 1979 imposed an asset charge of $15 to $18 or more a barrel, compared to only $1 in 1970. Payment of this charge imposes a serious economic burden on all countries that import oil. Their growth will be slower and the benefits of that growth will be smaller. Countries must devote more resources to intermediate goods and energy production and less to final products. Countries with oil—the U.S., USSR, Mexico, and others—may have better prospects. But for all oil-importing countries, the setback of 1974 and 1975 and the high and rising price of oil offer a less hopeful economic future.

In oil and energy, the years since the big oil-price increase have been a time of great soul-searching and study, but only modest actions. The high price of oil, the recession, and enforced conservation in most countries have cut the rate of growth of oil consumption. Industry, in particular, has reduced its oil-output ratio significantly. Motorists, however, continue along almost the same trend as in the past. Still, oil consumption, which in the past outpaced the growth of world production, now lags the increases in the world economy.

Oil is under the microscope, subjected to the most detailed studies everywhere. Strong oil policies, either to discourage consumption or to encourage production of oil or its substitutes, have been slow in coming, especially in the United States. Hundreds of studies are projecting oil supply and demand for the 1980s and 1990s, confirming oil's uncertain future. Indecision, however, assaults policy-makers and all manner of vested interests buffet them. The technologists are experimenting with many substitutes. Nearly everything remains on the drawing board.

Recent years have confirmed one fact. OPEC has irrevocably changed the distribution of world economic and political power. It has pushed its way forward with oil and dollars and has made a permanent niche for itself next to the great commercial nations. Saudi Arabia is now a major trading power. Iran, despite its revolution, remains a great Middle Eastern power. Nigeria is a leader in Africa, as is Venezuela in Latin America, and Indonesia in Asia. The exports of the thirteen OPEC members are greater than those of all of the less-developed countries together and exceed also those of either Germany or the United States.

In 1979 the tight oil market returned. The market jumped over the 14 percent price increase decreed by OPEC in December 1978, and members were adding surcharges early in the year. The June OPEC increase raised the Saudi Arabian price 42 percent above the 1978 price, and the price for the rest of OPEC was raised 57 percent over the 1978 price. Brisk demand assured that the market price would match or overrun these prices. Oil-importing countries bestirred themselves to agree on import quotas in an effort to stem further price advances. The end of the decade saw OPEC more concerned about oil deficits in the market than excess capacity but still firmly in control of the market.

OPEC ADJUSTMENT TO
THE REVOLUTION

With the new responsibilities of market command in 1974, OPEC worked hard. In addition to its regular conferences in June and December, OPEC also held four extraordinary meetings. Behind the scenes at the conferences, frantic activity on prices, the crude-oil pricing system, royalties and taxes, production, the condition of the market, and the use of the flood of money occupied officials of OPEC members. They set a pattern of peripatetic travel that still persists and of intense discussion and consultation, both in the OPEC headquarters in Vienna and in the capitals of members.

OPEC members were guessing when they increased the price in October 1973, by 70 percent. They guessed again when, only months later in January 1974, they pegged it at $11.65 a barrel, 3.9 times above the early October price. No one could predict whether or not the new price would so depress the amount demanded that a downward adjustment would soon be necessary. Indeed, some in OPEC thought an even higher price would not upset the market much more than it already was. The OPEC Conference in January 1974, called on the secretariat to recommend a new price for April. Conferences in June and September, however, reconfirmed the decision not to change the price.

By mid-1974 the strong market began to weaken. The cutback and embargo scares were over, and inventories of importers regained their former levels. The effect of the high price on consumers began to sink in. Surplusses began to clog storage facilities. The market price, which up to that time had hugged close to the OPEC price, started to show signs of declining. By September most OPEC members were cutting back production to strengthen the market price.

In November 1974 Saudi Arabia and other Persian Gulf exporters reduced their price by about 4 percent. Some others followed suit. The high OPEC price held, however, because nearly all OPEC members reduced their exports enough in the last half of the year to sustain it. In 1974 the average market price of $11.50 a barrel cleared the same amount of oil exports that an average price of $3.50 a barrel had cleared in 1973. By the end of 1974, the market was depressed and excess capacity was accumulating rapidly. Still, the oil market was functioning normally under full OPEC control.

After the 16-percent increase in exports in 1973, exports in 1974 remained constant. OPEC produced 30.8 million barrels a day in both years, and exported 29.4 million barrels a day in 1973 and 29.5 in 1974. In the last half of 1974, however, production for all members fell by 9 percent compared to the first half. The drop reflected the growing recession in industrial countries and importers' resistance to the higher priced oil.

The two largest importers, Saudi Arabia and Iran, who supplied nearly one-half of OPEC production, expanded in 1974 by 12 and 3 percent respectively. The only other members to expand, and by only small amounts, were Indonesia, Nigeria, and the United Arab Emirates. Production and exports of all other OPEC members declined. Two countries, Venezuela and Kuwait, bore the brunt of the cutback. Venezuela reduced exports by 21 percent. Kuwait reduced exports by 21 percent. In both cases the cutbacks were deliberate. Venezuela's weak reserve position impelled its actions. Kuwait's reserve position was in no danger. But neither had a pressing need for more money and calculated that keeping the oil in the ground was a good investment.

CASH AND CONTROL

OPEC members revenues benefited by more than the simple increase in price. Officials of OPEC members still believed that oil-company profits were extraordinarily high and indeed, that the companies had taken advantage of the oil crisis to fortify their profits. With excess capacity moving up, OPEC members, still greedy for more revenues but fearful of pressing consumers too hard, did not feel that they dared to squeeze another price increase out of the market.

Rather than increase the price, OPEC members decided to pressure the companies for more revenues. The government take of Saudi Arabian marker crude, which had been 99 cents in January 1971, and $3.00 in December 1973, was $9.27 for the first six months of 1974. For the next six months it was $9.37 a barrel, and in October, it was $9.69 a barrel. In the last two months of the year, it was

$10.69 a barrel. In four years government revenues per barrel had increased about ten times.

The increases in export earnings of OPEC members were staggering. In 1972 oil-exporting countries had earned $25 billion. In 1973 they earned $40 billion. Price increases accounted for about one-fourth of the jump that year, the increase in volume for about three-fourths. In 1974 OPEC members earned $120 billion. All of the increase in value over 1973 earnings came from the price hike. OPEC members had earned only 10 percent as much as the industrial countries in 1973 and 60 percent of the earnings of the other less-developed countries. In 1974 members earned nearly one-fourth as much as the industrial countries and nearly a quarter more than all the other less-developed countries.

Saudi Arabia was suddenly wealthy. Its exports earned $4.5 billion in 1973 and $31.0 billion in 1974. No country in the world has ever experienced a windfall of that magnitude in such a short period of time. But all OPEC members, after years of careful husbanding of foreign exchange, had money to burn. Venezuela had $3.1 billion in earnings in 1972, and had $10.8 billion in 1974; Iran had $4.1 billion in 1972 and $21.5 billion in 1974; Kuwait had $1.4 billion in 1972 and $7.0 billion in 1974. Saudi Arabian exports alone exceeded the exports of all countries in the world save seven: the United States, Germany, Great Britain, Japan, the Netherlands, Canada and France.

Having satiated its appetite for cash for the moment, OPEC members turned to greater control of the industry. Only a few months before the end of 1973, Kuwait obtained a 25-percent share of its companies, Gulf and British Petroleum. In the beginning of 1974, the share went to 60 percent, and later in the year Kuwait took 60 percent of Arabian Oil (Japanese) in the Neutral Zone. Saudi Arabia and Qatar were not far behind. In June the companies of these countries relinquished 60 percent of their holdings to these governments, retroactive to the beginning of the year. Nigeria acquired 55 percent of its companies. Other countries established sharing arrangements with new and existing companies.

The countries that had chosen the nationalization route also had an active year. Venezuela established a high-level commission to report to the president on how to nationalize its industry in an orderly way. Algeria took 51 percent of all new production companies. Libya took 81 percent of Occidental, nationalized 51 percent of Exxon and Mobil, and took over Royal Dutch-Shell's share of Oasis. The move to complete formal control or ownership was not complete in 1974, but through either participation or nationalization, OPEC members controlled most production. Decisionmaking about production and exports passed fully into OPEC hands.

Members of OPEC paid the oil companies for the properties nationalized or for the part ownership they received. The basic price was the net book value of the assets. The companies complained that this figure was too low and that replacement value would be more appropriate. Their position was weakened for having claimed high depreciation on their capital earlier in order to avoid taxes. In some cases, OPEC members did allow an inflation factor to be applied to the value of assets. Although the companies were unhappy and haggled over the price, most negotiations went smoothly.

OPEC'S SUMMIT MEETING

In March 1975, OPEC threw a party. Never before had such a strange group of countries paraded in all their finery with such pomp and ceremony. But never before had a group of countries achieved such a victory. The party, held in Algiers, toasted oil, OPEC, and the comeuppance of the rich industrial countries. It was a victory celebration.

Presidents and generals, chieftans and kings, sheikhs and the shah kneeled before oil, their savior and their hope. Well they might. It had yielded nearly $230,000 a minute the year before. Oil paid for the colorful robes, elegant uniforms, Saville Row suits, as well as the airplanes, shiny black cars, and great banquets. Oil paid for the armed forces that kept most of the potentates in power. And oil held the promise of economic development for their countries and of influence in world councils. "Noble oil," as the Shah of Iran called it, and cooperation, had given OPEC members a new place in the world.

Poor countries had held meetings before, but not poor countries that were rolling in money. When poor countries had met before, they complained of their plight, criticized the rich countries, and bickered among themselves. In this OPEC summit meeting, no diatribes against former colonial masters or neocolonialism stirred the delegates, no pleas for more trade and aid, no fretting about the multinational corporations. These countries had joined the establishment. They had muscled their way into partnership with the great nations in the management of the world. They felt and expressed the burden of responsibility that power had thrust upon them.

The Solemn Declaration at the end of the meeting intoned,

"The Kings and Presidents fully realize the close links between national development of their own respective countries and the economic prosperity of the world as a whole" (published by OPEC, 1975). Although the recession was in full swing in the industrial countries and members were cutting back exports, the bosses of OPEC shied at taking the blame. If the whole world would cooperate, as OPEC had, they argued, the world could banish inflation, recession, and underdevelopment.

Speech after speech emphasized cooperation, not only among the thriving thirteen, but also in the entire company of the poor and with the industrial countries. Oil would be the spearhead of change and the fount of progress. Those who have it must lead in the creation of a new international economic order. The new order, speakers argued, would remedy the deficiencies of the world economy and distribute the benefits of economic progress equitably. Poor countries had but to join in global cooperation to realize their potential.

Everything was sweetness and light on the podium. The Shah of Iran and Saddam Hussein of Iraq, whose countries were ancient rivals, kissed and made up. Conservative Sheikh Yamani and revolutionary Major Jalloud conversed, smiling and joking. But lurking in the back corridors was dissension, clashing of national interests, and distrust. The oil market was going down. For the camera, however, everyone smiled. The party was a huge success.

OPEC AND THE LESS-DEVELOPED COUNTRIES

One of the oddities of history occurred right after the oil-price revolution. An objective evaluation indicates that an anti-OPEC alliance between the oil-importing less-developed and developed countries would have served the interests of both. The disapproval and combined efforts of all oil importers might have restrained OPEC. The less-developed countries, hurt more by OPEC actions than the industrial countries, could have found ample justification for trying to strike back at OPEC. But they didn't.

Rather, OPEC and the less-developed countries formed an alliance. Leaders of less-developed countries, of course, were not overjoyed at paying $12 a barrel for oil. They may even have recorded some unkind remarks about OPEC in their diaries. But most of the less-developed countries were so elated that somebody had finally stuck it to the United States and the big industrial countries that they could bear their pain. They even cheered OPEC on.

The Third World did not strike the OPEC members from the membership list.

Even before OPEC's rise to power, the less-developed countries had discussed for some years reforms in the world economy to improve their position. Their cooperation—through the Group of 77, the United Nations Conference on Trade and Development, the United Nations General Assembly, and regional organizations—was not strong and their bargaining position was weak. But they were beginning to recognize that only their cooperative efforts would ever carry any influence and that raw-material prices were central.

The results of OPEC's command of the oil market was an example of what the less-developed countries wanted. Cooperation and control of a raw-materials market had paid off big for OPEC. OPEC was the success story of the less-developed countries. Now they must all go and do likewise. The real adversary for both OPEC and the less-developed countries, they reasoned, was the control of markets by the industrial countries. The new power of OPEC, hitched to the cause of the less-developed countries, stood at least a chance of wresting from the industrial countries some economic reforms.

OPEC members also had a choice. They could have aligned themselves with the industrial countries, entering into an advantageous producer-consumer commodity agreement. They could have stayed aloof, running their own affairs, including the world oil industry, to their own profit. Or, they could use their new power to help the less-developed countries to effect changes in world economic relations that would benefit them all. They chose the last.

At the preliminary North-South discussions, the great powers, still smarting from the high price of oil, wanted to talk only about oil. OPEC said that they would talk about the entire development problem, oil included, or nothing. Their position was firm. At the OPEC summit meeting in 1975, the sovereigns and heads of state had solemnly declared the unity of OPEC with the less-developed countries. Now the entire Third World wielded the oil weapon. The first meeting between the poor and the rich broke down. Soon the great powers came back to the table on OPEC terms.

The Paris North-South Dialogue was the beginning of bargaining over the more equitable sharing of world production. The delegation from the less-developed countries included seven OPEC members. Much heat warmed the hall and some light illuminated it. The industrial countries upheld the virtues of markets. The less-developed countries pointed out that one can't eat virtue. They argued that most markets confer most of their benefits on those who need them the least. No breakthrough occurred. Some say the meeting was a failure. It wasn't. It was just the first round, and it ended in a tie.

Not only did OPEC lend its political and diplomatic support to the less-developed countries, it also put its money where its mouth was. No sooner had the deluge of money started coming into OPEC hands than it began to flow out again. Every OPEC member started a foreign aid program. Most of them were larger, on a per capita income basis, than those of industrial countries. In 1974 OPEC recorded flows of nearly $6 billion, nearly 2 percent of their gross national product. The next year it was more than $8 billion, about 3.5 percent of their gross national product, ten times the ratio for the industrial countries.

THE 1975 DOWNTURN

Evidence that the January 1974 price increase was too much for the market to absorb steadily mounted. From mid-1974 through 1975 the market slumped, threatening to drop the OPEC price and forcing a large reduction in OPEC exports. The gap between what OPEC members did produce and what they could produce widened. The 1975 production declined, falling to 27.1 million barrels a day— a fall of 12 percent. Production in 1975 was only slightly above that of 1972. The Saudi Arabian cutback was the greatest. Iran, Venezuela, and Kuwait also suffered large losses. Iraq, alone among OPEC members, increased in exports. All members produced far below capacity.

The *Petroleum Intelligence Weekly* estimated that in 1975 OPEC members could have produced 39 percent more than they did. OPEC members could have produced 37.5 million barrels a day (mbd). Saudi Arabia, the largest exporter, had the most excess capacity. It produced only 7.1 mbd with facilities that could have yielded as much as 10.8 mbd. Kuwait operated at 62 percent of capacity. Venezuela, Iran, Libya, and Nigeria also had abundant excess capacity. OPEC was running on only five of its eight cylinders. If only a part of the extra 10.4 mbd had appeared on the market, the price would have plunged. It is a credit to the strength of OPEC and a small miracle that the price held.

Rather than sell all the oil they could produce for whatever the market would fetch, OPEC members chose to keep the oil in the ground. They would sell what they could at their fixed price, expecting to sell the difference later at yet a higher price. In effect, they invested in their own oil. The holdback meant earning a lower income than they would have earned if they had sold all they could produce at the existing price. OPEC members, however, estimated that they were earning a higher income than they would have earned if they had sold all they could produce at the price that would have prevailed with all supplies on the market. Suppose that they sold 10 percent more but in selling that much more, the price dropped 12 percent. They would have had fewer dollars.

Members would also have had less oil to sell later. With all the ups and downs in the market, OPEC estimated that the price trend must be upward. Only a fixed amount of oil exists in the earth. As supplies dwindle, the growing scarcity will eventually be reflected in the market price. By holding their oil in the ground they not only avoid the income loss that might result from high exports at a low price, but also still have the oil to market later at a higher price. Keeping their oil would earn more income for members on both counts.

To illustrate: OPEC sold 27.1 mbd and earned $110 billion in 1975. If they had produced 37.5 mbd and sold it at $7 a barrel, an optimistic price, they would have earned $96 billion, some $14 billion less. And they would have had 3.8 billion less barrels of oil, which if sold later at the 1975 average price of $11.10, would bring in $42 billion. Total loss to OPEC: $56 billion.

The first of two dilemmas emerged in the slump year. To keep the oil in the ground required that members restrain production. They had no production quota system. They had tried for years to install one but failed. Each member, keeping in mind its own best interests, had to decide voluntarily how much oil to produce. If all of them decided to produce an amount that totalled what the market could absorb without breaking the price, then all was well. If they produced too much, the market price would fall below the OPEC price, hurting everybody. The pressure on the large producers—Saudi Arabia, Iran, Kuwait, and Venezuela—who had to accept a reduced market share, was immense. In 1975 they held together.

The slow oil market and its high price placed OPEC members in another bind. Consider the oil-price problem in 1975 from the OPEC perspective. These thirteen countries had achieved a great economic victory with the price increases of 1973 and 1974. In 1974 alone, however, the export prices of the United States—the prices OPEC members had to pay to buy American products— advanced 27 percent. Japanese export prices moved up 28 percent and German export prices, 19 percent. OPEC members helplessly watched their bonanza shrivel in their hands as the prices they paid for imported goods increased.

In 1974 OPEC members defended themselves in part by raising royalty and tax payments, squeezing oil-company profits without raising prices. In 1975 the high price and recession depressed the oil market. In addition, more and more oil-company facilities came under full OPEC control, reducing the option of squeezing profits. The inflation continued, down somewhat from 1974 because of the recession, but still significant. American and German export prices rose 12 percent. British prices advanced 16 percent. Furthermore, OPEC members began to complain that the prices they paid for imports were higher than those paid by other importers. Every week a barrel of OPEC oil would buy less; therefore, revenues were down, and the value of financial reserves shrank.

What could OPEC do to stem the tide? The only remedy left was to increase the price again. OPEC members knew that they risked reducing their exports even more, adding to their excess capacity. They could only hope that the worst of the recession was over and that a small increase in the oil price would not pinch off very much demand. The growth of the economies of oil-importing countries could rescue them, they hoped, preventing a further serious fall in exports.

Long and hard debates marked the 1975 OPEC Conference in Vienna. Some members wanted a large increase in price, matching, since 1974, the inflation in the industrial countries. Indeed, some members suggested an automatic arrangement—indexation—in which crude-oil prices would move up with an index that reflected increases in the OPEC import price index.

Other members wanted only a small increase. They feared that a large increase in 1975, in a bad market, would worsen the recession in oil-importing countries and further stiffen consumer resistance. Even more excess capacity might accumulate. The excess capacity would then be a temptation to OPEC members to cheat and threaten the price. Voluntary restraint would break down. So sensitive was Saudi Arabia to these issues, that it wanted no price change at all. In the end the members compromised. At the beginning of the fourth quarter of 1975, OPEC increased the price of crude oil by 10 percent.

Despite the large drop in production and exports in 1975, OPEC members still earned $110 billion that year, only $10 billion less than the year before. Every country except Iraq and the United Arab Emirates lost revenues that year. Saudi Arabia's earnings from exports were down by $3.3 billion. Venezuela lost $2.0 billion. Iran, Kuwait, Libya, and Nigeria each lost $1.0 billion or more. Still, compared to 1973 earnings, OPEC members were in clover.

The 1975 downturn did not damage OPEC greatly. They salvaged some income by raising prices. The greatest impact—one-third of the income loss—fell on the one country, Saudi Arabia, that could sustain the loss most readily. Ample financial reserves fortified its ability to cope with the loss. The income of all members in 1975 was four and one-half times greater than in 1972. The gross national product of OPEC members remained the same in 1975 as in 1974. Most of all, however, the bad year proved that OPEC could stick together, even when the market declined.

THE 1976 RECOVERY

Production of crude oil by OPEC members bounced back in 1976. Members produced 30.6 million barrels a day, less than 1 percent below the record highs of 1973 and 1974, and 13 percent greater than those of 1975. Venezuela's production declined, but only slightly. Libya expanded 31 percent, and Saudi Arabia, 21 percent. Iran, however, the second largest producer, increased production only 10 percent. Saudi Arabia emerged clearly as the dominant OPEC member, with 28 percent of OPEC production and 15 percent of world production. Still, in 1976, all members were producing below capacity.

The price debate within OPEC continued. The price set in October 1975, did not change during the year. At the OPEC meeting in Bali, Indonesia, in October 1976, the Saudi Arabian dominance became obvious. Despite improving production and earnings, as well as the decline of inflation in industrial countries, most OPEC members wanted a price increase. Iran, whose rate of expansion in 1975 was less than half that of Saudi Arabia, was especially vociferous. But Saudi Arabia said no and that was that. Trouble was brewing, however, as the conflict between members whose revenue needs compared to earnings were great and those, like Saudi Arabia, who were accumulating vast financial reserves.

OPEC earnings in 1976 jumped to $135 billion. The Saudi Arabian share was 27 percent, with earnings up by $8.4 billion. Iran was second, but its earnings increased only $3.3 billion. Three other countries—Iraq, Kuwait, and Venezuela—each earned $9.3 billion or more, and Indonesia, Libya, and the United Arab Emirates each earned $8.4 billion or more. In the three years 1974-1976, OPEC members earned $365 billion, more than twice what they had earned in the decade of 1964-1973.

The price freeze in Bali had steadied and improved the market. In December 1976, OPEC met in Doha, Qatar, amid speculation of another price increase. The usual controversy between the price hawks, led by Iran, and the price doves, led by Saudi Arabia, titillated the reporters at the conference. But this time, the argument within the conference was more profound and the hawks challenged Saudi Arabia. The less-affluent OPEC members hungered for more revenues.

TWO-TIER PRICING

Before the Doha meeting Sheikh Yamani said "the world economy wouldn't tolerate the effect of a strong rise at this time. Even an increase of 5 percent in the price of petroleum would slow the recovery of the industrial countries" (*Wall Street Journal*, December 15, 1976, p. 1). In addition, for Saudi Arabia, another new element entered into the equation. A new and unknown man was soon to become President of the United States. What better way to earn his gratitude and to test his intentions in the Middle East than to show restraint in oil prices? If Saudi Arabia kept the price down, President Jimmy Carter would take office owing Arabs a debt.

Iran and other members that were hurting for money were unimpressed with Sheikh Yamani's concern for the world economy. His game of one-upmanship with President-elect Carter also left them cold. They pointed to the signs of recovery in the oil market. They showed how the continued price increases in their imports were damaging them. They wanted more money now. Dr. Amouzegar, speaking for many members, had a hold-out minimum of 15 percent. Sheikh Yamani's maximum was 5 percent. In the past they had always been able to compromise. This time they couldn't.

No unanimous decision on a single OPEC price came out of the Doha Conference. Instead, the members unanimously agreed to disagree. Nominally, they preserved unanimity; in fact, they did not. The Venezuelan oil minister, Valentin Hernandez Acosta, devised the face-saving formula. Dr. Amouzegar of Iran and ten other members got their 15 percent, but in two stages. The price for eleven members would advance 10 percent at the beginning of 1977 and another 5 percent in the middle of 1977. The other two—Saudi Arabia and the United Arab Emirates—opted for a 5 percent increase at the beginning of the year and a reevaluation of the price at mid-year.

Iran and the members who had increased prices 10 percent noisily warned Saudi Arabia not to permit its exports to expand at their expense. Sheikh Yamani, sitting on more than $40 billion in international reserves, reserves second only to those of Germany, smiled and reminded his colleagues that Saudi Arabia did not control the market. Importers buy from whom they please. He placed the majority under tremendous pressure to hold their prices down. One, Indonesia, joined the lonesome two, and others, nominally retaining the higher price in fact sold for less. Sheikh Yamani reminded President-elect Carter of the importance to Saudi Arabia of a quick and comprehensive Middle East solution and of his country's role in keeping the oil price down.

THE 1977-1979 MARKET

Despite the price increases of 1977, OPEC members produced a record amount of oil: 31.4 million barrels a day, an increase of 2 percent over the year before. Still, the market was not prosperous. Toward the end of 1977 and through most of 1978 the market went sideways. Indeed, in much of 1978 the market slumped, recovering only late in the year because of Iranian cutbacks. Production for the year was 29.9 million barrels a day, about 5 percent less than in 1977 and also less than 1976. The abundance of oil and downward pressure on prices in these two years led many in the oil-importing countries to believe that the oil problem was over.

The two-tier pricing of 1977 caused some shifting of market shares among OPEC members. Production in Iran declined 7 percent. All of this loss, however, is not attributable to the price differential. Late in 1977 strikes in the oil fields and the growing revolution took its toll. Saudi Arabia expanded by 7 percent. Production and exports for most other members remained about the same, as total OPEC exports increased only 1.1 percent. The Saudi Arabian position in 1977 would have been even stronger had it not been for the setback occasioned by an oil field fire.

The market shifts, noticeable in changing revenues early in the year, created intolerable strains. Two prices, 5 percent apart, for roughly the same grade of oil, was enough to induce oil companies to buy from Saudi Arabia and slow down liftings from Iran. Saudi Arabia dealt from a strong hand but could not ignore the other members without jeopardizing OPEC and inviting retaliation. All through the winter and spring months the price discussions among members continued. Finally, as the deadline for another 5 percent increase for the higher tier countries approached, those members decided to forego the additional increase.

A part of the deal that led the high-price countries to restrain themselves was the Saudi Arabian agreement to raise its prices by 5 percent more. By midyear, OPEC reunited at a price for 1977 that was 10 percent more. Two-tier pricing ended. The disagreement was a severe test for OPEC. Experts in industrial countries were overjoyed and some foresaw the breakup of OPEC. Ali M. Jaidah, secretary general of OPEC told me (Interview, *Worldview*, May 1979, p. 42) that the dispute was not the beginning of the end of OPEC, but rather positive proof of its strength. His reasoning, as it turned out, was correct. But it does not forestall other disagreements within OPEC.

OPEC earnings in 1977 set another record, increasing to nearly $150 billion, a 9-percent increase. Saudi Arabia alone gobbled up almost two-fifths of the increase because of its lower price for half of the year. The value of its exports began to rival those of Canada, Italy, and the Netherlands. Iran was disappointed with its 3-percent increase. But still, it earned more than $24 billion, more than twice that of any Middle Eastern country except Saudi Arabia. Six of the other OPEC members had exports close to $10 billion. These earnings made these countries major world trading and financial powers.

Late in 1977, although production increased in anticipation of a possible price increase at the beginning of 1978, the oil market began to sag. This put pressure on members not to increase the price at the OPEC conference in Caracas in December 1977. The old dilemma of a stable oil price and rising import prices emerged. Inflation in the industrial countries had heated up. The U.S. export prices advanced 9 percent; Japanese 17 percent; and German 13 percent in 1977. In addition, to add to OPEC woes, the value of the dollar skidded to new all-time lows. OPEC had to decide whether to increase price to offset inflation and compensate for currency losses or to keep the price stable for another year.

Three positions surfaced at the Caracas meeting. Most members wanted a modest increase—5 to 10 percent—to improve their position without the disruption of a large increase. The three radical Arab members—Algeria, Iraq, and Libya—wanted an increase of one-fourth, arguing that only such a price increase would preserve the purchasing power of their earnings. The idea of indexation came up once again. The practicalities of the moment persuaded most members that a more moderate course was desirable. Saudi Arabia posted the third position—no price increase at all.

Iran, usually eager for a price increase, did an about-face. It sided with Saudi Arabia in arguing for no increase. The Shah feared that a higher price might jeopardize his revenues and open up the controversy with Saudi Arabia which had brought them to grief earlier in the year. In addition, the Shah had visited Washington shortly before the Caracas meeting. President Carter had encouraged the Shah to restrain his enthusiasm for high prices. The United States also agreed to support some of his defense needs and to supply nuclear reactors for his long-term energy needs. With the two largest members urging stability, the OPEC Conference adjourned without taking any action on price.

Most of 1978 saw the oil market continue in the doldrums. Some months, especially early in the year, exports were back down to 1975 levels or below. Alaskan and North Sea production created a temporary abundance in most oil markets. Some members shaved the price a bit to keep their markets, but generally the OPEC price held under pressure. The mid-year OPEC conference again did not raise the price, fearing further accumulation of excess capacity. After mid-year the market picked up a bit. With Iran's production falling at year's end, Saudi Arabia and others increased production, and all of OPEC were exporting at record levels.

Members continued to complain loudly and bitterly about losses from inflation and currency depreciation. Regularly, the press reported that this or that OPEC member asserted that his country was losing millions or billions of dollars, that a price increase would be necessary, and that it may be necessary to abandon the dollar. Members talked about a price increase all through the fall of 1978. At the Abu Dhabi Conference in December, they raised the price for 1979 by 10 percent.

The initial 1979 price increase installed the staged method OPEC had long considered. On January 1 the price would advance 5.0 percent; on April 1 by 3.8 percent more; on 1 July by 2.3 percent; and on October 1 by 2.7 percent. Taking $12.70 as the December 31 OPEC price, the price for October 1 would be $14.54 a barrel. The price increase amounts to an average increase for the entire year of 10 percent.

Although OPEC pricing is simple in concept, in operation it is complex. The actual price increases do not match the announced price increases. This results from the way the complex oil market functions, with many sub-markets and intricate company-country relations. The OPEC posted price for Janaury 1, 1974 was $11.65. The OPEC price at the end of 1978 was $12.70 a barrel, an increase of 9 percent, although announced price increases by OPEC in that period were double that increase.

No sooner had OPEC announced the new 1979 prices than it became clear that Iran was dropping out of the market. It had exported 4.9 million barrels a day in 1977 and 5.2 million barrels a day in 1976. It was OPEC's second largest producer, and in early 1979 its exports went to zero. Saudi Arabia took up some of the slack, as did other OPEC members, but not enough to make up the shortfall. After a few months, Iran began to export again, but much less than it had previously. Add the fall in exports to increased demand in early 1979, occasioned by continued growth in importing countries and especially because the United States miscalculated its import needs, and the market went into a tizzy.

Early in 1979 the staged method of pricing was abandoned by OPEC and all members went at once to the October 1 price of $14.54, an increase of 14.4 percent. Then members began adding surcharges, largely because they found that importers were eager to pay more to get the oil. By mid-year, the market price stood at about $17.50, 37.8 percent over the 1978 price. In late June, 1979, OPEC changed its prices again. Saudi Arabia, in company with the United Arab Emirates and Qatar, raised its price to $18.00. For the rest, the price went to $20.00 a barrel. Members agreed on a ceiling of $23.50 a barrel on contract prices. The oil market was in such turmoil that no one was certain that the new OPEC prices would be enough to restrain demand.

The largest oil-importing countries, meeting shortly after the OPEC Conference, deplored the OPEC price action but failed to note that most of the increase simply brought OPEC prices into conformity with existing market prices. For the first time, however, oil importers, instead of just complaining and wringing their hands, agreed to set oil import quotas. These quotas will not halt the growth of imports completely, but they do indicate a willingness to cooperate in restraining demand. For OPEC, the 1979 market ratified their decisions and actions through the decade and conformed their control of the oil market.

OPEC WORRIES

For all their progress and great influence at the moment, OPEC members worry about world conditions beyond their control. Chief among these worries are the purchasing power of oil and the value of their financial assets and their earnings. From the beginning of 1974 through 1978 the price of their oil had increased only 9 percent while the index of world export prices generally had increased more than 2.5 times. If the price of oil had kept pace with other export prices, its price at the end of 1978 would have been about $30 instead of less than $13.

Since oil is priced in dollars and payment to OPEC is made in dollars, the deterioration in the value of the dollar also damages OPEC. When members buy German or Japanese goods, their dollars buy fewer marks and yen, which means an added price for imported goods from those countries. Their financial assets held in dollars suffer the same fate. In addition, the high rate of inflation eats away most of the earnings of the dollar assets. OPEC members feel that these double vises, both held in place by the United States and other industrial countries, has slowed their economic progress.

The other major problem that OPEC worries about is the demand for crude oil. With modest ups and downs the market has stagnated since 1973 despite the economic growth of oil-importing countries. The benefits to OPEC have come almost exclusively from price increases, especially the 1974 increase. The reasons for the oil market's failure to advance stem, of course, primarily from OPEC's own price actions. Until early 1979 OPEC had substantial excess capacity that was always a threat to OPEC unity and to its price.

Paradoxically, OPEC worries about a market that is too prosperous. The rate of production is so high that each year a substantial chunk of reserves disappears. At the end of 1977, OPEC proven oil reserves were 440 billion barrels, only slightly above their 1970 reserves. If the rate of production, 31.4 million barrels a day, or 11.5 billion barrels a year, continued for thirty-eight years, present proven reserves would be gone.

OPEC members are aware, of course, that more oil will be found. But they also know that consumption will increase. The increase in price from $3.01 a barrel to $12.70 a barrel at the end of 1978— 4.2 times—permitted a slight expansion of consumption. Most of the projections of demand show increasing consumption in the 1980s and 1990s. So OPEC worries about perilously low reserves and high rates of consumption that would endanger not only the oil market but also threaten their prospects for development.

OPEC feels that it cannot cut prices because the lower price would not encourage enough more exports to offset the price drop. Hence, OPEC income would go down. Most of the oil-exporting countries, at least the dominant members of OPEC, feel also that they cannot raise the price in a slow market. A higher price might accelerate substitution, stimulate oil-importing countries to stronger conservation policies, induce a recession, and depress demand further. To defend their incomes, some OPEC members might then try to enlarge their share of the market by price cutting. But only by raising the price can OPEC members offset some of the effects of inflation and the declining value of the dollar.

OPEC members have considered squirming out from under the dollar by shifting to a basket of currencies. They would retain the dollar as the means of payment, but the price of oil, the number of dollars paid per barrel, would be set by currencies, including the dollar, that both appreciated and depreciated. If the dollar deteriorated further, the price would remain stable since the other currencies would be gaining as the dollar lost value. During earlier dollar troubles between 1971 and 1973, they tried a basket and it worked. Still, so far members have been unwilling to cut loose from the currency of their best customer and the strongest currency in the world. Their dilemma remains.

Two niggling problems bother OPEC. The OPEC conferences have become almost exclusively price-bargaining sessions. Indeed, so much time is devoted to the price problem, that many other problems of the industry receive scant attention. Many members are seeking ways to reduce the time dedicated to price, although they are unwilling to give up full participation in the price decision. One possibility is setting the price only once a year, leaving the other conferences free for other industry problems.

The OPEC method of determining price had also imparted a characteristic movement to the oil market. In the two months before each OPEC Conference, demand picks up, to a greater or lesser degree, depending on whether or not buyers expect a price increase. Buyers try to build inventories just before the price increase. In other periods, demand flags. The staged price increases for 1979 attempted to address the problem. The tight market resulting from the Iranian revolution upset their plans. Indeed, no orderly price procedure works well in a bull market. OPEC's price is a minimum price, sustained by holding back exports. But market forces drive the price above the minimum when demand booms. OPEC can increase exports but cannot regulate price in a seller's market.

Still, OPEC's complaints are minor. In the period 1969 through 1973 these thirteen countries exported 40.8 billion barrels of oil and earned $117 billion. In 1974 alone they earned as much, exporting 10.0 billion barrels. In the period 1974 through 1978 OPEC members exported 48.5 billion barrels for a return of about $635 billion. The five-year windfall from OPEC cooperation was $500 billion.

Before examining how OPEC has used this greatest windfall in the history of international finance, it is necessary to see how world oil markets operate and what differences their control by OPEC has made. These matters occupy the next chapter.

Chapter 6

The World Crude-oil Market

The oil market is really two markets. One is the market for oil products. It is governed by supply and demand. The other is the crude-oil market. The demands of refiners of products determine the demand for crude oil. Earth's bounty and the enterprises or countries that control crude oil determine its supply. Prices in both markets still respond to how much users of products want to buy and how much those who control crude oil want to sell.

Neither the oil companies nor OPEC members have repealed the law of supply and demand. The great oil companies, through their control of crude supplies until the 1970s and their influence over the demand for products, did bend the law and make the market for oil products and for crude oil serve their purposes. OPEC, through its dominance of supply and willingness to accept the consequences of the demand for products, has created a new crude-oil market that since 1973 has served its goals.

The consequences for the price and the amount of crude oil differ. When the companies ruled the roost, the price was low and stable. Production of crude oil expanded rapidly to meet the demand for oil products that the companies manipulated to serve their interests. With OPEC in control, the price is high and rising because that market behavior serves its members' interests. Production of

crude oil expands slowly, conforming to the amount that the product market can absorb when the OPEC crude-oil price is included
in the price of the products.

The differing motives and goals of the companies and OPEC
members explain much of the differences between the past and
present oil markets. These differences consist not only of price
and production and of their rates of change, but also of the changing
supply and demand conditions that OPEC dominance has wrought.
In addition, the OPEC market has altered the market's institutional
arrangements and has brought into being new institutions and
methods.

When Exxon, British Petroleum, and the others, including the
independents, made the market decisions, they sought to achieve
three goals—profits, stability, and growth. They used economic and
political means to pursue these goals. Their great economic power,
relative to that of the producing countries, enabled the companies
to influence political decision-making. Political decisions then
added to profits, stability, and growth, which in turn augmented
the political influence of the companies. The companies did not
seek political power for its own sake. Politics was only a means.
Economic gain was the end.

As sovereign nations, the members of OPEC necessarily pursue
political as well as economic goals. All of the companies pursued
nearly identical goals and cooperated to achieve them. OPEC members' goals are more diverse and complex, and they sometimes
clash. The crude-oil industry is now a political as well as economic
entity. The oil companies, still in command of most of the demand
side of the crude-oil market, continue to pursue their economic
goals. But they are bereft of most of their political influence in
producing countries and are subject to the countries' supply decisions. In politicizing their decisionmaking, the members of OPEC
not only created a new crude market, but also a different kind of
market.

The OPEC market has brought into sharp relief the difference
between the short and long-run oil markets. The short-run market,
in which OPEC now fixes the price and accepts the demand that
comes forward at that price, responds to the transitory and changing conditions of the moment. The long-run market embodies
only the persistent trends in the short-term markets, as well as
capacity, reserves, and ultimately the amount of the nonrenewable
resource. The short-run market may result in either declining production and downward pressure on price, or just the reverse. In
either case, at that moment, in the long-run market the scarcity

value of oil inexorably moves up under the pressure of reduced supplies and expanded consumption.

Among economists and politicians in developed countries it has become popular to call OPEC a cartel. The meaning of the word is cloudy but its meaning to those who use it is pejorative. Whether or not OPEC is a cartel depends on how a cartel is defined. Most economists define a cartel as an agreement among large enterprises controlling most of the production of a product to set prices or production in order to maximize the profits of those in the agreement. OPEC clearly controls the price of crude oil through decisions on supply. Its dominance of price, however, is not immune to influence by the demand side of the market and its control of supply is incomplete. Political and strategic matters, concern for the world economy, and multiple and differing goals of members attenuate OPEC's pursuit of the economic gain that is the overriding goal of cartels. Hence, the results differ.

THE MARKET FOR CRUDE OIL

The market for crude oil, or indeed for any good, is like a pair of scissors. One blade is demand. The other is supply. Like scissors, markets function only when the two blades interact. Demand does not determine the price or amount consumers buy any more than one blade of the scissors cuts paper. The market too, does not set the price and fix how much is sold without demand. Markets relate the price, the quantity that people want to buy, and the amount that producers want to sell. The market operates when price moves up or down to equate what consumers want to buy and what producers want to sell.

No one really wants to buy crude oil. It is a black gooey stuff, not much good for anything in itself. The demand is for products made from refined crude oil. Motorists want gasoline, light oil, and lubricants. Householders want fuel oil. Utilities want heavy oils to make the electricity. Industry wants products for heat and power and as feedstock for petrochemicals. The demand for crude oil comes from the markets for oil products. When these product markets establish prices and producers know the amounts of products they can sell at various prices, then the buyers of crude oil—the refiners—know how much they can offer for crude at different prices.

Most users of oil products want to buy nearly the same amount regardless of the price. The motorist wants thirty gallons a week whether the price is 60 or 90 cents a gallon. Technology often

dictates how much oil products industrial users must buy. An industrial plant buys the same amount of oil products if it is producing the
same amount of output regardless of price. Economists call this kind
of demand inelastic, meaning that the quantity that consumers will
buy is relatively unresponsive to price changes.

Inelastic demand has a special meaning for sellers. If the price
goes down, not enough new demand appears to offset the drop in
price. Sellers as a whole earn less money. If the price goes up, the
increased price discourages so little demand that the earnings of all
sellers go up. These effects apply only to sellers as a whole. An
individual seller may fare differently. In addition, there are many
degrees of unresponsiveness to price changes. At one extreme, a
price change does not affect demand at all. In the middle, a price
change is matched by an equivalent change in amount, so that
sellers earn the same amount. On the other end of the scale, a price
change may result in a large change in the amount demanded so that
when price goes up, the sellers' income goes down, but when price
does down, the sellers' income goes up.

Many products have inelastic demand. Products that are a necessity or that people regard as a necessity can have widely varying
prices with but little change in the amount people want to buy.
Products that have no handy substitute, as well as some that are
only a small proportion of the user's total expenditures, share the
same fate. Most agricultural products have inelastic demands. This
explains why farmers all over the world lobby for higher prices
and appeal to their governments for help when prices fall. This
also explains why OPEC wants a high price.

The oil-products markets impart to the crude-oil market their
lack of price responsiveness. An increase in the price of crude may
result in some belt tightening, but refiners have learned that when
they pass along the higher crude price, users of products will buy
about the same amount. So refiners will want to buy about the
same amount of crude oil even when its price changes. When, for
example, the price of crude went from an average of $3.50 a barrel
in 1973 to about $11.50 a barrel in 1974, buyers bought the same
amount of crude oil.

Over a period of time consumers are more responsive to price
changes. After a price increase has been in effect for some time, the
change sinks in and the consumer rethinks his attitude. The motorist decides that he can forego his Sunday drive and the next year he
may buy a smaller car or take his vacation some place closer to home
than Yosemite. The industrial consumer, who tends to be more
price conscious than the individual, installs equipment using coal

instead of fuel oil. He also makes adjustment in whatever oil-burning equipment he must keep in order to economize, and he lowers the temperature in his offices and warehouses. The year after the price increase, OPEC exports declined and the market has since been slow.

More than just price influences the demand for oil products and hence, for crude oil. The incomes of consumers and the production of agriculture, industry, and commerce play a role. With a raise in pay, the worker buys a second car, and with higher production, the farmer buys another tractor. When gross national product goes up, users buy more oil products as they produce more of their own product, even though the price remains the same. Should production go down, as it did in most countries in 1975, consumers buy less oil products. Rising incomes since 1975 help to explain the modest increases in OPEC crude-oil exports.

The price and availability of substitutes also influences oil demand. If a price increase puts the oil price above that of coal, natural gas, or other forms of energy, then users will switch, buying more of the substitute and less oil. It takes time and the substitute's price must be low enough, compared to oil, to compensate the user for the cost of switching. One of the reasons Europe has been so successful in conserving oil is that many countries now use a great deal more natural gas. New technology, as it comes along in the use of oil or in machines using energy, will also influence oil demand.

Government policies often influence the demand for oil products. In 1973 the U.S. government mandated a 55-mile-per-hour speed limit and has subsequently set automobile fuel efficiency standards. It has also required that utilities expand their physical facilities with coal-burning equipment and continues to push nuclear and other forms of energy. Governments can also influence the price of oil products and through the price, their demand. Nearly all of the oil-importing countries tax oil products heavily. In Europe, for example, crude oil is only about 30 percent of the cost of oil products. Taxes take 45 cents of the oil-product dollar and the oil companies get the remaining 25 cents. In the United States crude oil is a higher proportion of product costs, but taxes are a much smaller share of the product price. A change of 10 percent in the price of crude is a change of only 3 or 4 percent in the price of oil products if taxes and company charges remain the same.

In the ideal market, competition brings supply and demand together at a unique price. If the price departs from that single market-clearing price, competition among the sellers, in the case of a surplus in the market, will bring the price down. In so doing, consumers are enticed by the lower-than-market price to buy more, and some sellers are squeezed out of the market because their costs are too high to make a profit at the lower price. The surplus disappears.

If a market shortage prevails because of a low price, competition among buyers will force the price up. At the higher price, some consumers drop out of the market, and some producers enter the market with expanded production. The shortage evaporates. Oil-product markets have many imperfections, especially when large firms can influence their own supply and demand conditions. Most product markets however are reasonably competitive. This makes the demand side of the crude-oil market competitive.

Even though a relatively small number of firms, including such giants as Exxon and Royal Dutch-Shell, buy most of the crude oil, they act as intermediaries for millions who do compete to buy oil products. Their desire to make money serving consumers and the competitive fringe created by many small companies compels all firms to compete vigorously in buying crude oil. Until 1973 the large enterprises that bought crude oil were also the sellers of crude oil. They often controlled the sources of crude oil. Exxon was thus a seller in the product market and both a buyer and seller in the crude-oil market.

Indeed, for much of the crude oil an open market barely existed. When Exxon sold to Exxon, the price that the company charged itself was a bookkeeping entry, not a real price that said something about supply and demand. Although all oil products go through markets, only a portion of the crude oil went through the crude-oil market. Some companies, even large ones, could not fill their product needs from their own crude sources. Other companies had more crude than they needed. A crude-oil market existed although legal contracts regulated most of the transactions. Only a very small spot market existed. The large companies set the price of crude oil but they were influenced by the bargaining that culminated in company-to-company contracts and by the volatile spot market that reflected mainly immediate supply and demand conditions.

Since 1973 OPEC members have controlled crude-oil supplies and have set the price. Most of the oil continues to go through the same channels, even the same companies. The oil contracts are now negotiated between governments or their national oil companies and the buying private oil companies and the bargaining centers around small deviations from the announced OPEC price. The spot market also continues and as before, influences the OPEC minimum price and the prices established in contracts.

OIL-COMPANY SUPPLY AND PRICE

Economists have special words to describe the methods that firms use in interacting competitively as suppliers. At one extreme, competition—often called pure or perfect competition—prevails when no firm can influence the price at which it sells its products because there are so many firms. It is rare. Most often, economists use it only as a model and measuring rod against which to compare most realistic forms of competition. Monopolistic competition is common. These competitors do have some influence over their price, especially over the price of their own brand, which is differentiated slightly from the others. Still, monopolistic competitors face the competition from the other firms that produce similar products.

A few producers—two to twenty, or even more—can control prices, often influence their own demand, and interact in unpredictable ways. This is called oligopoly, and it is most common in resource (except agriculture) and manufacturing industries. Indeed, most energy, metals, and manufactured goods in the United States and many other industrial countries are produced under oligopolistic conditions. The extreme of oligopoly is the monopoly, also a common market form. The monopolist controls supply and sets the price and behaves predictably. Because of their great power, many monopolies, such as the utilities, are regulated.

In the crude-oil market, the large oil companies were oligopolists. Seven firms controlled most of the crude-oil supplies. Other firms existed and at various times, such as in the 1950s and 1960s, had considerable influence. But the big firms called the tune most of the time. The unusual aspect of oligopoly is that because there are only a few firms, each watches the others like a hawk and conditions its own behavior not only on what the others do, but also on what it thinks the others will do. For example, Exxon, in considering a price increase, may decide not to raise its price because it thinks that Royal Dutch-Shell won't, leaving Exxon high and dry. On another occasion, Texaco, in thinking about a price cut, may forego it, fearing that others will follow suit.

All firms, making the same sort of assessment as Exxon and Texaco, also decide to do nothing. The price remains stable. But this is only one possibility. The interaction of decisionmaking among a few large enterprises ranges all the way from cutthroat competition to all the firms acting as though they were one firm. One large firm may try again and again to drive his competitors out of business, as did the old Standard Oil, with predatory pricing and market penetration. The other firms respond in kind. The industry is in constant turmoil. But four or five firms can, by accurately estimating what other firms will do, present a solid and stable front, acting as though they were just one giant enterprise.

Economists usually assume that all enterprises try to make as much money as possible. The technical term is profit maximization. In pure competition, the result of the assumption of pure competition is zero profits for all firms. In monopolistic competition, the same assumption also yields zero profits, but with costs that are higher than would have been the case in pure competition. In oligopoly, costs are also higher and there can be persistent profits, or varying profits and losses, depending on the interaction among the firms. Monopolists can make profits.

The trouble with this analysis is that many enterprises seek something other than maximum profits. Henry Simon, the 1978 Nobel prize winner, has shown that enterprises, especially the large ones, are very complex and follow many muses. John Kenneth Galbraith, that needler of the orthodox, argues that growth to most firms is more important than profits. Most economists, he asserts, sacrifice accuracy in the interest of the simplicity of the profit-maximization assumption.

The desire for stability and the growth of the enterprise may indeed supercede its desire for profits. When a large firm is subject to the predacious actions—price cutting, aggressive advertising, market penetration—of other large firms, it longs for the comfort in which its survival is assured and it can plan its future with confidence. When the market grows and the other firms are growing, a firm feels that it must also grow, at least as fast as the others, else it is losing ground. To be sure, firms need profits, but they will often trade some of their profits for greater stability and growth.

Large enterprises have an aversion to shifting market shares, with large shares one year and slim the next, to uncertainty, and to high risk. The oil companies conducted multi-billion dollar enterprises. They needed to plan their finances and make investments in new fields and refineries, in technology, expansion, new products, and in other matters far into the future. Ups and downs, fighting to achieve or retain control of product markets or crude oil, and unsettlement over what is going to happen next were anathema. Large enterprises would prefer a profit formula, that is, a target for profits, rather than maximum profits, if with some profits they could buy peace.

Growth is the real success indicator for most large enterprises. If others grow faster, a firm is a failure, even though it may have huge profits. For the management, the company's growth guarantees good jobs, increasing responsibilities, enlarged prospects for more pay, benefits, prestige, and the feeling, for its executives, that they are with a winner. Growth also protects management from maurading outsiders who would like to grab control of the firm. Those who run the firm do not benefit from maximizing profits, but they do benefit directly from growth and stability. Stockholders, an ignorant and uncaring lot in most enterprises, will hold still for less than the highest profits if the company is safe, stable, and making progress.

The triad of goals—profits, stability, and growth—governed the behavior of the oil companies. Other motives sometimes intruded for short periods under special circumstances. These motives determined the interaction among the firms as well as production and prices. After the long period of instability and uncertain growth and profits in the early days of oil, the large companies formed a cartel in 1928. This cartel was an agreement among the oil oligopolists to divide up the market among themselves in a fashion that assured each of them greater profits, stability, and growth. The Achnacarry Agreement was a formal "hands off" agreement in which each large company gave written assurance to all the others that it would not engage in the earlier predatory practices and would, from that time on, leave the markets and crude sources of the others alone.

The cartel did not last. The passage of time and changing circumstances undermined it. New oil fields, new products, and new markets altered market shares. The companies also discovered that they could achieve most of the results of the cartel simply by behaving so as not to incur the wrath of the other companies. If each company correctly estimates the reaction of the other companies to its actions, it can avoid the penalties that the cartel was established to prevent. Among the large companies, a live-and-let-live policy followed the disappearance of the cartel. It became second nature and natural. The cartel died but its spirit lived on.

The result of this organization of suppliers differs from competitive industry strikingly. The economic power conferred by great size, steady growth, stability, and adequate profits enabled the companies to influence and often to control much of their own environment. They were able to secure from governments favorable tax treatment, actions that promoted the sale of their products, and access to cheap raw materials. When the managements were meeting all their goals, they had sufficient power to keep the stockholders, labor unions, and governments happy—and to keep outsiders from interfering.

One of the most important benefits of the firms' success in achieving their goals was the ability to influence demand. According to the economist, enterprises cannot influence demand under competitive conditions. The oil companies, however, learned that they could influence demand directly. They created demand and expanded it through technology, government, advertising, and mutual back-scratching with other industries. A steady stream of new, redesigned, and repackaged products, balleyhooed by all the media, stimulated the consumer to buy more and more. The growth of such things as complementary industries, automobiles and highways (thanks to the government) benefitted the oil companies.

The oil companies helped to persuade the government to build the superhighways, to encourage the building of single homes, and to stimulate lenient credit. Consumers could hardly wait to take long vacations in their gas-guzzling cars, to burn clean fuel oil to keep their homes above 72 degrees, and to spend next year's incomes to buy oil products. The power of the companies and associated industries was so great that they could reach into the psyche of consumers and convince them that they must behave so that only an ever increasing amount of oil products could yield the good life.

To do this required that the companies keep the prices of oil products low and production high and rising. They needed the low price to complement their manipulation of consumers to buy more. Then they adjusted production to meet their customers' expanding needs, which the companies had helped to foster. They tried to keep the government from taxing oil products. But even their great power could not prevent the taxation of products so easily subject to taxes.

Above all, the oil companies had to keep the cost of crude oil low and production responsive to the needs of oil-product customers. They were able to do this for a long period of time. Through long-term concessions obtained under favorable circumstances, modest royalties, and low taxes in the producing countries, the companies could pass the low price of crude oil along to consumers. Had the producing countries been in a more favorable bargaining position from the 1920s on, they could have earned much higher incomes at that time by making consumers pay higher prices for oil products. Producing countries didn't have the power, and it was in the companies' interests to serve the consumers.

Let there be no doubt—the oil companies served consumers well. They provided huge amounts of oil products at low prices and consumers counted themselves better off. Industry was able to make goods cheaper because of cheap oil. A significant part of the expansion of the industrial countries and the welfare of their peoples can be attributed to the availability and low price of oil before 1973. A significant part of the continued stagnation of oil-producing countries—Mexico, Iran, Venezuela, and many others—must be attributed to their powerlessness and the consequent low crude-oil price before 1973. Did the price reflect accurately competitive supply and demand conditions for crude oil? Or did it result in the arbitrary transfer of income from oil-producing to oil-consuming countries, carried out through the exercise of the economic power of the oil companies?

Most of the costs paid by the oil companies were straightforward. They had to pay for the concessions. They had to undertake geological studies and explore for oil. When they found it, they had to drill wells and get the oil above ground. They had to transport it and store it while awaiting export. The companies had to pay the taxes and royalties required by the country that owned the oil. They had to pay all the costs associated with converting the crude into products and getting the products into the hands of consumers.

This list omits one cost, the cost of the crude oil itself. The companies regarded crude oil under the ground as a free good, theirs for the taking. They paid only the costs of extracting the oil and getting it to market. Yet crude oil is a scarce substance, having a value independent of production costs. It is also one of earth's fixed assets, exhaustible and not reproducible. Neither the companies nor economists, however, know its true value.

In the economist's ideal world, land has value and the owner of the land receives a rent for its use. By extension, a resource, such as crude oil, has value and its owner should receive a payment that reflects its value. But the economist's competitive world departs so greatly from the real world in which power, technology, the past, present, and future, lack of knowledge, and market imperfections play such determinant roles. No way exists to compute, in theory or practice, crude oil's real value.

The oil market does not help much in establishing crude oil's value. The crude-oil market is derived from the oil-product market. In the latter market, supply and demand determine the price. But part of the supply side of the product market is determined by the value—cost—of the crude oil. The cost of crude oil, however, does not take into account the value of the crude oil itself. Crude-oil costs do include payments to the owners but these payments are far from the economist's notion of the cost of crude oil.

As the surrogate cost, the companies have always regarded the concession, royalty, and tax payments to the governments of oil-producing countries as their crude-oil cost. The payments were necessary in order to acquire the right to extract and export the crude oil and to transfer legal ownership of the oil to the companies.

The payments to the producing countries, however, bore no relationship to the supply conditions of crude oil or to the fixed amount of oil remaining, either in the country or in the world. The companies did make the payments, however, and passed them along as costs, but not as costs that reflected the scarcity value of crude oil.

Concession payments were a matter of bargaining between the companies and the governments of oil-producing countries. The companies, always anxious to keep the payment low, took advantage of their superior bargaining position that they possessed at the time. The oil cartel and agreements, such as the Red Line Agreement, often gave a single company or only a few companies exclusive rights to seek concessions in a given country before the 1950s. Rarely was there competition among companies to obtain concessions that resulted in increasing the income of the producing country significantly. But even when a country bargained astutely, the result had no relation to its supply of crude oil.

The producing countries were poor, needed any amount of money badly, knew and cared nothing about oil, and had weak, often corrupt, governments. They often settled for a token payment. The amount may have seemed large at that time and may indeed have been large compared to the existing revenues of the country. But the payment was trivial compared to the value of the oil. Since the amount of oil the country possessed was unknown, concession payments could not reflect the value of the country's deposits.

The royalty was the share of crude-oil production received by the producing country for the alienation of its patrimony. Taxes, mainly based on the income of the oil companies, also reflected the amount of production, not the scarcity of oil. In both cases, the companies paid in money. They had to assign a price to the crude. The companies paid according to their posted price, the price the companies used as an indicator of the market price, which was in turn determined by the market for oil products. They endeavored to keep the crude price low and stable to support their efforts to encourage oil's use and to keep production high and rising. The low price also kept payments to the oil-producing countries low.

When the members of OPEC took control of price and production in 1973, they included a more generous allowance for themselves in the price of crude oil. The amount of the government take per barrel went up about fifteen times between 1970 and early 1979. But the new price of crude, including the new government take, or asset charge, is no less arbitrary than the old price when the companies paid modest concession, royalty, and tax charges to the producing countries.

The companies, in order to achieve their goals, kept the cost and price of crude oil low when they had the power. OPEC has insisted on a higher price and has the power to make its decision stick. In both cases, power determined the price, the power of companies cooperating with consumers and the power of the owners cooperating with the companies. In neither case does the arbitrary price of crude oil reflect its scarcity value. OPEC does argue, however, that its approximation to the real value—the OPEC price—is more appropriate to an era in which oil is becoming increasingly scarce.

OPEC SUPPLY AND PRICE

The OPEC takeover of the world crude-oil industry didn't happen in a day. True, on a day, October 16, 1973, OPEC set its own price for the first time. But OPEC had had that power for a year or more before that date. Some members had been setting their own posted prices. Venezuela established its own tax-reference price and Libya posted its price. Others could have. The power of OPEC members to determine price originated in their ability to determine, on their own, the royalty and tax rate, and to determine the amount of production.

Libya had forced the companies in 1970 to negotiate a higher price. The Persian Gulf producers had also negotiated a higher price under threat of a crude shutdown in 1971. In the next two and one-half years OPEC bargained with the companies on the price. On October 8, 1973, OPEC members sat at the price-bargaining table with the companies. When the meeting failed to establish a new price, OPEC members a week and a day later sat at its own price-bargaining table without the companies and negotiated a new price among themselves.

Production control by producing countries stretches over an even longer period. Iran's nationalization, which gave that country some limited say in production, goes back to 1951. In the late 1960s and early 1970s, several members nationalized their companies. Just before the price takeover, Saudi Arabia and Persian Gulf Arab countries negotiated partial ownership of their companies. In 1974 the process continued, with more countries nationalizing or taking ownership of companies. Formal control or ownership is still not complete. Some companies continue to operate concessions as before or have arrangements with OPEC members that vest them with operational control. Even so, the companies must obey the dictates of members on pain of takeover.

As production and price control passed to OPEC members by many different methods and at different times in the early 1970s, they had many options. They could operate as a unit, coordinating their decisions, or each member could go its own way. Having achieved success by operating as a unit, they decided to continue. As a unit, they could establish a formal cartel and accept its logic, or bargain and coordinate decisions among themselves in a less formal way. Since members were unwilling to give up any national authority, they chose bargaining and informal coordination.

In the oil market, OPEC members also had two options. They could fix prices and let production vary, or they could fix production and let price vary. Under the first option, the market would determine production for each member. Under the second, the market would determine the price at which each member sells. The members could not fix both production and price at the same time. The market acts like the fulcrum of a lever. Apply force to one end and the other must move in a predetermined manner. OPEC members chose to set a minimum price and let the market determine the amount of production.

OPEC members could have benefitted more economically by establishing a cartel and behaving like a profit-maximizing monopolist. They were not quite a monopoly, but 85 percent of production was sufficient to behave as one. To have behaved as a monopoly, the members would have to set production at a level so that if they produced less, costs would go down less than sales, or if they produced more, costs would have increased more than sales. It would have been difficult, since costs differ among the members. They could have, however, used revenues as the profit indicator.

OPEC chose not to establish a cartel and behave as a monopolist. Members were unwilling to give up their right to decide how much to produce. A monopoly of several units producing the same product must agree on the shares each will produce. OPEC had tried throughout the 1960s to agree on a sharing plan but failed. When they had nearly complete control of production, they still couldn't agree. In addition, to maximize profits required fine calculations of revenues, profits, costs, and exports, and a central organization to administer and police the system, all agreed upon in advance. They couldn't agree. Indeed, OPEC members could not agree on any method involving setting production that removed decisions from national authorities. Centralized production control of any kind was out.

OPEC members were accustomed, however, to having the price

dictated. The oil companies had been doing it for years. When the shift of control took place, it was only natural that instead of the companies dictating price, OPEC would dictate its own price, through their oil ministers. They would agree on a price among themselves—a minimum price. Then each member would sell as much as it could at that price, without any price competition. The meeting of October 16, 1973, became the model for future OPEC operations. The oil ministers would bargain among themselves, decide unanimously, set the price, and then accept the production that the price entailed. The veto preserved national authority. Later, the oil ministers would meet again, assess any changes that had take place in the interim, and set another price.

Another possible option for OPEC members would be to permit both price and production to vary, but at different times. OPEC could set the price and then let the market determine production. For a period of time OPEC would then maintain that level of production. If demand changed, the price would rise or fall with production fixed. Still later, the members would set the price again, allowing the market to fix production. Over a period of time the members could play back and forth between price and production, fixing one, then the other, to achieve their goals.

The oil companies alternately fixed price and production for a long time with considerable success in an expanding market. Although cooperation among the oil companies was never complete and independents often disrupted the plans of the large companies, the majors would often set rates of production expansion with stable prices. When the rate of expansion threatened a world oil glut, the companies used falling prices, which eased their cost position, to maintain production.

Cooperation among OPEC members, also imperfect, has yielded changes in both price and production. In the surplus market from 1975 through most of 1978, OPEC members tended to produce too much. In some cases, prices fell. In others, the announced price increases of members were never attained. Later, in the short market, especially in 1979, the market price rose much higher than the announced OPEC price and production increases were insufficient to restrain price.

Neither the companies nor OPEC have a rigid price and production policy. The companies tried to keep the crude-oil price low and production high and rising. OPEC tries to keep the crude-oil price high and rising, with production and production increases sufficient only to maintain prices in a glut and push prices up when demand is strong. Above all, OPEC members want to retain control of price, which means that they must cooperate not only with respect to price but production as well.

The companies, of course, did not disappear when OPEC took over. Although many were partially or wholly nationalized and the ownership of others, especially those in the Persian Gulf, passed into members' hands, most of the companies continued to pump oil. There was no grand exodus as happened in Mexico in 1938. In Iran, of course, in 1978-1979, not only foreign oil workers, but foreign workers in all sectors of the economy departed in a wave of anti-foreign sentiment. In all OPEC members, executives were replaced by nationals, but many professionals, technicians, and skilled workers remained, working for the host government under the new arrangements. Some of them have gradually left, as their counterparts have taken over their jobs. New national oil companies and special enterprises appeared as the governments of OPEC members moved to full control.

When OPEC members took over operating companies in the field, the home offices of those companies often established new special-purpose enterprises. These new companies provided services to members' governments or their national oil companies. The services included technical services, buying and marketing the members' oil, geological and exploration work, and sometimes even operating services. In other cases, those in which the operating companies remained, government regulations and contracts specified the rights and obligations of the companies and the government.

Indeed, relations between the private oil companies and the governments of OPEC members have improved greatly since the OPEC takeover. Members realize that the companies can perform many functions that they cannot. They particularly need the technology and technical assistance of the companies. They also require assistance in management and sometimes, in operations. Members need the companies for equipment and for field maintenance, as well as for marketing and transporting the crude. Without reluctance, OPEC members use the private companies and pay them well for their services.

The difference is that OPEC members now feel that they are in control. The companies are there at the member's request, not by right. The members pay the companies as they would any employee or enterprise and direct their activities. The companies, once trust is established, are often given great leeway. Ultimate control, however, rests with the OPEC member. The companies are not disturbed by this arrangement. They are well paid for doing the things that they know how to do well. As national personnel in OPEC countries learn the tricks of the trade, some of these services will drop out. But both the companies and OPEC members expect a long-term and mutually profitable association.

A sticky point in nationalization and participation has been company compensation. The principle adopted by OPEC specifies that the companies will receive net book value of the company assets. The companies assert that their assets are worth much more and they have the support of their home governments. The result in most cases is that a negotiated settlement begins at net book value and goes up. Members often include an inflation allowance, an adjustment for claimed assets in the past, and a different valuation for particular assets. But OPEC refuses to accept the concept of replacement value, urged by the companies. Although the companies have not been happy with their settlements, they have not been sufficiently unhappy to permit a dispute over compensation to interfere with their relations with member governments.

Five methods govern most OPEC-company relations. In the most common commercial transaction, a national oil company of a member contracts with a representative of a private oil company for the sale of oil. Another method is the production sharing arrangement. The oil company uses its own capital to find oil and takes all the risks. When the oil is found and production begins, the company covers its costs and then shares, by contract, the oil with the member. Another method is the risk contract. The company finances the search for oil. If it finds oil, the government reimburses the company for all of its costs, but then takes all the oil. If the company doesn't find oil, the company is out.

Yet another method is the service contract. The service contract shifts the risk to the government, which pays the company a fee for exploring, operating a field or a refinery, or providing specified services, such as technological aid. The service contracts differ widely in their provisions. Still another method is the joint venture. In effect, the OPEC member and the company become partners. In these arrangements, the companies and OPEC members share ownership, risk, and responsibility.

These arrangements have been amicable and sufficiently advantageous to OPEC members that many no longer feel compelled to insist on complete ownership. OPEC members' control is complete. Using these methods or variations on them, the members can take advantage of the superior capabilities of the private enterprises without losing control. Nearly everything is done by short-term contract, often highly detailed and specific, so that both parties have methods to measure the benefits and costs and have the opportunity to renegotiate in a short period, if dissatisfied. These arrangements have been so successful that members frequently use the oil companies, as well as other foreign private enterprises, in many aspects of their general development programs.

Nearly all OPEC members have established a national oil company. They have a bewildering array of interesting acronyms: SONATRACH, Societé Nationale pour la Recherche, la Production, le Transport, La Transformation, et la Commercialisation des Hydrocarbures in Algeria; NIOC, National Iranian Oil Company; QGPC, the Qatar General Petroleum Company; KNPC, Kuwait National Petroleum Company; PETROVEN, Petróleos de Venezuela; PETROMIN, the General Petroleum and Mineral Organization of Saudi Arabia; and many others. Some countries have two or more public oil companies.

The national oil companies range from operating enterprises to holding companies. Some have only partial control of the industry in the country, sharing with private foreign oil companies and other national oil companies. Others have complete control from exploration to refined products. PETROVEN, for example, owns the twelve operating companies that took over from the private companies under the 1976 nationalization and it makes all industry-wide decisions. Some companies are little more than offices in the oil ministry, completely under the dominance of the government. Others have great autonomy and often dictate policy. All, of course, are public enterprises and are ultimately subservient to the government.

OPEC members still sell most of their oil through the international oil companies. The oil companies who operate in the member nation, a special buying enterprise, or a resident representative contracts with the national oil company or the ministry to take delivery of a specified amount of oil per day for a period of time—30 to 180 days in most cases—at an agreed price formula. The mechanisms of pricing are extremely complicated, depending on the grade and other physical characteristics, seasonal market conditions, charges of the companies, payments arrangements, and many other matters. Pricing for refinery production is even more complicated.

OPEC members do not sell all their oil to the companies that formerly controlled the oil. Although members frequently grant such companies a preferred buying status, assuring the company of oil and the member of a market, most members endeavor to develop their own direct sales. Members also sell to intermediaries. A part of OPEC oil finds its way into the spot market, a small but noisy market whose supply is usually already in a tanker bound for the market abroad. Companies and intermediaries buy and sell this oil, sometimes more than once. The spot market price is very sensitive to short-run supply and demand conditions and often is an indicator of the direction and strength of price movements.

Most oil moves at figures close to the OPEC price. Bear in mind that the OPEC price is only for the market grade of Arab light. The OPEC price for that oil was $13.3390 a barrel on January 1, 1979. At the same time, some still lighter grades were selling for up to $1.40 a barrel more and heavy grades for 50 to 90 cents less. Nor are the differentials constant, but change as supply and demand conditions change. In addition, OPEC members in a sluggish market often discount their oil slightly, as they did between 1975 and 1978. In late 1978 and early 1979, on the other hand, they charged a premium above the announced OPEC price.

In setting the OPEC price, members seem to have only the most elementary strategy. That strategy calls for a high price, a rising price, and price control by OPEC members. Each members has a different notion of what the price should be, and as time passes, each member alters its views depending on the state of the market, its own supply situation, and many other economic and political considerations. The views of the members differ greatly. Each OPEC conference witnesses a battle royal over the price to be set for the ensuing period and over the framework of pricing. It is usually an amicable but intense altercation that ends in a unanimous decision to raise the price by a given percentage, or to leave it alone.

The OPEC semiannual conferences have come to be price-bargaining meetings. Thirteen positions are possible but usually only three emerge. The price doves—Saudi Arabia and the Persian Gulf Arab states—want no increase in price or only a modest one. The price hawks—Algeria, Iraq, Libya (the most extreme) and Iran (more moderate)—want a large increase. Most of the other members are somewhere in between, usually closer to the doves than to the hawks. Some of the hawks want to tie the price of oil to the prices of goods they must import. They also want to substitute a basket of currencies for the present method of pricing in dollars. The loss in value of the dollar and continued inflation provide ample fuel for impassioned appeals for justice and complaints against the industrial countries, who by conspiracy or ineptness, perpetuate inequity.

All members of OPEC are conscious of the technical, statistical, and administrative difficulties of indexation—tying oil prices to import prices—and pricing in a basket of currencies. Only a few want to risk the experiment. Often hawkish members use these ancillary issues more as an argument to influence the doves and those nations with dove-like tendencies to raise the price more than is their inclination. Most members do not want to install any automatic pricing system that removes decisions on prices from the bargaining table.

Until late 1978 and early 1979, the condition of the market and the course of international events have supported the price doves. With no method for determining market shares, the variable and sluggish market of 1974-1978 placed a heavy burden on voluntary production restraint. Saudi Arabia, Kuwait, and Venezuela bore the brunt of voluntary cutbacks in production. The last two, however, reduced production primarily to conserve reserves. Saudi Arabia, by virtue of its unique position as the largest producer with the largest oil and financial reserves, is the fall guy of the voluntary restraint system. It is also the one invariable price dove.

The price conflict goes on and on, in conference and out of conference. Officials of OPEC members periodically make comments that may or may not reveal their positions in the closed OPEC meetings, but are designed rather to influence their colleagues, as well as the industrial countries. Even Saudi Arabia, with a chorus of support from others, tries to use the price-increase warnings to cudgel the industrial countries into restraining inflation and the United States into doing something about the dollar. Inflation and the dollar, of course, are only two of the many items that members consider when making a decision. The oil market, markets for other products, the condition of the world's economies, and all political and economic circumstances and events play a role in OPEC pricing.

The OPEC conference itself has become an element in the oil market. Before each meeting, the market is upset. Since price is on the agenda, buyers, to one degree or another, become convinced that the price is going to go up. The result is a surge in demand in the weeks and months immediately preceding the conference. Following the meeting, whether the price went up or not, demand falls off because buyers' inventories are high. Demand peaks twice a year, adding to the normal seasonal demand swings. The result is that for much of the year, the OPEC oil industry operates with excess capacity.

At the Abu Dhabi meeting in December 1978, members tried to break this pattern. They set price increases for January, April, and October all at the same time. It was the intention of members to establish a new pattern of annual pricing by quarters announced once a year. The plan blew up almost immediately. In late December 1978, exports from Iran halted. Other members, notably Saudi Arabia, made up some of the 5-mbd loss. Still, the market suddenly became very tight. In a short time the market price stood well above the OPEC minimum and most members went to the October 1 price, a 14.4 percent increase over the previous December. Some also charged a premium over that price. The market still churned with spot prices going to $25 a barrel or more.

The conferences have come to be concerned primarily with price matters. Yet OPEC has many other matters to consider, including downstream operations, long-run price and production strategy, the methods of pricing, the functioning of national oil companies, and many other matters that the secretariat has brought to the attention of the oil ministers. The ministers want to limit price bargaining to one session a year, leaving the other meeting for their other affairs. Some members have even suggested that price bargaining be delegated to a subcommittee or to a meeting of the ministers' representatives, freeing time for other discussions.

OPEC members, in company with most of the oil experts and seers of the future of oil, believe that the oil market since 1974 has been changing to one of increasingly limited supplies and continuing increases in demand. No one expects a return to the booming market before 1973. But still the possibility exists that in the 1980s, demand will consistently run ahead of supply, becoming chronic and acute at some time in the next two decades. Some OPEC members are speculating that if this permanent tight market emerges, it may be to their benefit to shift from bargaining over price to bargaining over production, leaving the market to set the price.

If the members set production limits in a tight market, the price may move up by enough to provide for their revenue needs. Such an organization of the market would also give members more control over the rate of depletion of their reserves, a matter of increasing concern to many members. OPEC as a whole has proven reserves sufficient for forty more years at the rate of production of late 1978. Most members feel their oil reserves are inadequate, unless husbanded with great care, to last even through the initial stages of the development of their economies.

The change to production-oriented decisions by OPEC is still in the talking stage. Members are uneasy over the present method— price setting—because it confers on buyers the decision about the rate of production and makes OPEC members the world's residual suppliers, bossed around by the whims of buyers and oil availability outside of OPEC. Ideally, they would like to control both price and production. They realize they cannot. They also realize that controlling production would suit their purposes only in a rising market that assured continued price increases of sufficient magnitude to provide the needed reserves, as the Secretary General has indicated (Interview, *Worldview*, May 1979, p. 42). So, until the market develops a chronic long-term shortage, members will bargain over price.

OPEC members are convinced that their market is more efficient than that of the oil companies in bringing the short-term and long-term markets together. By stabilizing the price at a high level, OPEC applies steady pressure on buyers to conserve oil. A variable market—one that encourages buyers to believe that the price may go down soon—does not permanently discourage the extravagant use of oil. In view of existing and projected supply and demand conditions, OPEC believes that its high price is closer to the true scarcity value of oil and encourages the needed conservation and the technological development of substitutes that will replace oil in the next century. The substitutes will provide the link between the short-term market and the long-term market, not only of oil, but also of all energy forms.

IS OPEC A CARTEL?

For a hundred years, the word cartel has signified a combination of private businesses that through the control of supply and price can make more money through cooperation than they can independently. Economists have analyzed them and governments have considered them illegal. Enterprises participating guard the secret of the cartel's existence closely. These private enterprises, subject to legal action by one or more countries with the physical assets hostage, endeavor to forestall government actions that might disrupt the arrangement.

Is OPEC a cartel? This is a matter of semantics. It is not the traditional cartel among private firms. It is cooperation among governments to control the operations of an industry they own or control through dominance of supply and price. It does not have the attributes of the historical cartels and often does not behave as the cartels of economic history have. But if economists and politicians want to call whatever OPEC is a cartel, then only the English language suffers.

The principal feature that distinguishes OPEC from the traditional cartel or an ordinary oligopoly is that its members are nations and its activities are confined to their own territory. As such, OPEC members are answerable only to themselves in the handling of their domestic resources. Industrial countries may regard OPEC actions as unfair, just as OPEC members have regarded the actions of industrial countries and their companies as unfair for decades. But OPEC cannot be illegal. No international law covers such an organization. Its members may do whatever they please for whatever reasons tickle their fancy. They may behave as any economic person. They may pursue political and strategic goals. They may act out of pique or make irrational decisions. Those decisions have limitations, political and economic, but the limits far transcend those of a cartel whose members owe their existence to national authorities.

Some reason that since OPEC is a producer cartel, then a consumer cartel could effectively negate its strength. This notion formed part of the basis for creating the International Energy Agency and other efforts at consumer cooperation, all of which have yielded disappointing results so far. The reason is that OPEC offers none of the handles which sovereign power can grasp and which nations traditionally employ against cartels. There are no OPEC producing assets, inventories, citizenship, charters, permits, legal entitites, or privileges in consuming nations. The only way consumers can influence OPEC is through the market, by reducing demand. Even that is ineffective unless it is of such a great magnitude that it exacerbates OPEC's internal tensions so much that OPEC falters from within.

The antitrust suit brought against OPEC by the International Association of Machinists in the United States is a quixotic gesture that involves potential hazards. It is improbable that the union could win a court case against sovereign nations. If the union did win and tried to attach the financial assets of the members of OPEC in the United States, financial chaos could result. Members would divest themselves of all financial holdings in the United States, a blow to the orderly operations of financial markets. In such a bizarre eventuality, other countries might also refuse to hold financial assets in litigious countries for fear of similar treatment.

Economists are a bit uneasy about labeling OPEC a cartel. They do so often because they are unable to find its proper classification. One well-known oil economist—the American John Blair in *The Control of Oil* (Pantheon, 1976)—has said that the world oil industry is now a symbiotic bilateral oligopoly, a new bird in economics. Two oligopolies, OPEC and the oil companies, cooperate with one another in controlling the industry. Another possible description: OPEC is a collusive oligopoly of public enterprises operating only in their own territory that cooperate with offshore buyers.

The urge to call OPEC a cartel comes from the need to predict its behavior and fate. If OPEC is a cartel, then the economists can pull the history books off their shelves and study them, along with the latest books on economic theory, to discover what OPEC will do and what is in store for it and its market. When the economists do this, they find a comforting fact that they want to believe. They find that all cartels collapse. They find the historical evidence and the analytical reasons. If all cartels have collapsed and for good reasons, and OPEC is a cartel, then OPEC will collapse.

The analytical reasons, given the economist's assumptions, are compelling. All enterprises want to make as much money as possible. That is their over-riding goal, indeed, their only goal according to the economist. The enterprises of the cartel can make higher profits as members of the cartel than they can independently. This works fine and holds the cartel together until one of the members discovers that if he secretly produces a bit more than he has agreed to in the cartel agreement, he can sell it at the cartel price or only a little below it. He can make even more money as a secret cheating cartel member than he can make as an honest member. Another member makes the same discovery. Soon, the cheaters must shave the price quite a bit to sell their added production. Still, the agreement-breakers are getting the benefits of the cartel—production restraint by the honest members—without restraining their own production as agreed.

The renegade members are taking income away from other members. Things like this can't be kept secret. The affected losing members notice it quickly in their income and production statements. A little looking around the market uncovers the reason. Someone isn't holding up his end of the bargain. To meet the threat, the honest members become dishonest too, cutting the price more to sell enough production to maintain their profits. In no time at all, the cartel members are competing with one another and the cartel no longer exists.

The historical evidence is equally compelling. In the past hundred years dozens of cartels have lived short tortuous lives in oil, chemicals, plastics, pharmaceuticals, explosives, and many other industries. John Eckbo, in his book, *The Future of World Oil* (Ballinger, 1976), examined many cartels and discovered that most live six or seven years. He dated OPEC's demise in 1979-1980. Cartels have waxed and waned but no cartel has maintained its control of supply and price for any lengthy period.

The facts and the logic are impeccable. Yet the conclusions are wrong, or at least highly debatable. It all starts with the debatable premise that OPEC is a cartel. If it is not a cartel or not the kind of cartel that history has experienced and that analysis has examined, then all that follows is wrong. The model of the cartel that OPEC is supposed to be requires that members always seek to maximize their profits. This one motive drives all the enterprises—countries, in the OPEC case—and accounts for all their behavior as producers.

If in fact cartels have sought other goals, such as stability and growth, then the collapse of the historical cartels must have come from something else. That cartels historically disappear is correct. They do so, however, not because they want to maximize profits, which makes them all dishonest, but rather because of public pressure, the disappearance of the cause that gave birth to the cartel, and the discovery of other ways to handle the problems to which the cartel was addressed. Cartels often serve a special purpose in the historical development of the industry that existed only at a particular time. When the circumstances of the industry changed, the cartel fell apart, having outlived its usefulness. In most cases, the enterprises found other and less formal means to accomplish the same goals as the cartel.

Motives other than profits undermine the reasoning and upset the historical precedents. For many large firms the desire for profits is tempered by an even stronger hunger for stability and thirst for growth. The private oil cartel displayed these mixed motives. The oil companies formed the secret cartel at Achnacarry in 1928 in pursuit not so much of profits as of stability. They wanted to eliminate the cutthroat competition and disorderly and uncertain markets that had existed in the industry in the previous thirty years. They also wanted to expand their production steadily. They also wanted profits. Many other cartels originated from these same motives.

The oil cartel did not collapse because its members cheated. To be sure, some did cheat, but that didn't bring it down. The cartel served its purpose of stabilizing the market, assuring growth for all, and supplying ample profits. Then the oil world changed. The predatory tactics of the earlier era were, by the 1940s, not a part of the thinking of the great companies and their executives. The whole industry had expanded, with the Middle East emerging as its center. The companies had learned to live with one another. The cartel was no longer needed. Companies behaved without it. Furthermore, it was risky. The possibility of antitrust action in the United States mounted. By the time the Federal Trade Commission released John Blair's study on the international oil cartel in 1952 and the Justice Department had readied its case, the cartel was a shadow of its former self.

The companies continued to cooperate without a formal cartel. As knowledgeable men in the oil industry, executives of the various large companies could estimate the behavior of other companies, as others could estimate theirs. They shared common attitudes and goals, knew what was best for the industry and for their own companies, and had no intention of upsetting the applecart for a momentary gain. They needed no cartel. The independents in the 1950s and many outsiders to the industry did not share these goals and caused trouble. But they were a small part of the industry. With the passage of time, some of the largest of the independents, with an increasing vested interest in the industry, began to behave like the majors. Other cartels have a similar life history.

If OPEC collapses it will not be because it is a cartel. So far, members have regarded the integrity of OPEC as one of their highest goals. Indeed, to date, OPEC unity has been the one supreme goal, above profits and all else. Members will suffer instability, forego economic gain, and accept zero growth or losses, if tolerating them will keep OPEC together and strong. They are acutely conscious of their situation before the success of their cooperation and have no intention to revert to those bad old days.

One of the overriding reasons for the emphasis on the integrity of OPEC is that members intend to use their cooperation to pursue a whole battery of goals other than profits. Money is obviously the powerful centripetal force. Without it there would be no point to OPEC. But since OPEC does exist and has yielded profits, members propose to use it also for all kinds of other purposes. Some of these goals are national; some are international. Some are unique to a single member; others are shared by many members. As nations, members have many interests and goals—economic, political, military, and strategic—and they can use OPEC to assist in their achievement.

Suppose that OPEC as a formal institution did collapse? What difference would it make? Would the price of oil plummet to former levels? Would the world go back to the pre-1973 situation? No. To be sure, there would be monentary disturbances in the oil market and the price might fall somewhat. Saudi Arabia, however, would still be the largest oil producer, followed by Iran (when it recovers), Kuwait, Venezuela, Nigeria, and Libya. Saudi Arabia would probably set the price of crude or would collaborate with one or more other countries in setting the price. Other oil-exporting countries, in their own self-interest, would accept the price. Discounting might be more frequent, the market might be more disorderly, but in the main, the oil industry would go on much as it does today.

All of the oil-exporting countries, including some, such as Mexico, that are not now members of OPEC, would find it in their interest not to compete among themselves through unrestrained production and price-cutting behavior. Norway, Mexico, and others have already indicated that they intend to pursue strict conservationist practices, code words for production restrictions. Informal cooperation and national behavior in the best interests of all would replace formal OPEC cooperation. All members would lose something, economically and politically, including the sense of a failed experiment that the world had predicted could not happen—a rematch of David and Goliath. But OPEC-like behavior and attitudes would survive the demise of the formal OPEC.

Backing off from the idea of a cartel somewhat, some economists have called OPEC a dominant-firm oligopoly. The sources of the belief are statements such as "No action is taken in OPEC unless Saudi Arabia wants it," as the *Oil and Gas Journal* quoted Sheikh Yamani (July 19, 1976, p. 102). Thirteen nations are members but ten produce 97 percent of exports and one, Saudi Arabia, produces nearly one-third. All interact and make agreements as would a collusive oligopoly. But the weight of Saudi Arabia is so great that, in fact, OPEC does its bidding despite the formalities that give each country one vote and require unanimity. So, the argument goes, OPEC is virtually a Saudi Arabian organization that functions to amplify that country's influence. Saudi Arabia is the power and OPEC the shadow.

This is a comforting doctrine. It shifts attention from OPEC— thirteen diverse nations—to Saudi Arabia, with more easily identifiable goals and capabilities. It also removes the problem from economics, since although Saudi Arabia has economic interests, it is a political animal. OPEC, on the other hand, is an economic animal whose members have political goals. Indeed, in this light, OPEC doesn't count and can safely be ignored as a separate entity since it is in the shadow of Saudi Arabia. This approach greatly simplifies the problem facing oil-importing countries since now there is a more visible and understandable target of policies. International diplomacy and economic warfare have methods for dealing with upstart nations basking in a moment of glory.

As in the case of the cartel characterization, the dominant-firm oligopoly applied to OPEC is an illusion. Economists' models have no history, no memories, no people, no subtle human interactions.

When OPEC was founded Saudi Arabia produced only 17 percent of OPEC oil, surpassed by Iraq and Venezuela. Venezuela's production doubled that of Saudi Arabia. Not until the early 1970s did Saudi Arabia emerge as an export leader and even then ran neck and neck with Iran for several years. For more than half the life of OPEC, Saudi Arabia was just one of the boys.

Libya, not Saudi Arabia, took the initiatives in 1970 that started the OPEC ball rolling. Iran, the radical Arab countries, and sometimes Venezuela are the price hawks and adventuresome members. Only with great reluctance did Saudi Arabia participate in the 1973 Arab embargo. The king of Saudi Arabia fired his oil minister—Abdullah Tariki—in the early 1960s because Sheikh Tariki wanted to move too far too fast. Saudi Arabia is a member of OPEC, an important member and its conservative leader, but it is still only a member.

Saudi Arabia alone could not have captured the oil industry in 1973. It produced only one-fourth of OPEC oil, only slightly more than Iran. In early October 1973, negotiations with the companies, Iran and the Arab states were the most intransigent and their behavior led to the breakdown of talks. Those negotiations had been fortified by the September OPEC Conference in which many members, but not Saudi Arabia, favored OPEC unilateral price setting. Saudi Arabia participated in these events and its added influence made the oil-price revolution possible, but Iran's voice was the most strident. If any country tried to behave as the dominant firm, it was Iran. But that country, without Saudi Arabia and the other members, could not dominate OPEC.

Saudi Arabia can, and knows that it can, force other members either to accede to its will or permit it to act independently. It forced a price compromise in 1977 and again in 1979, but still accepted higher prices than it wanted. Saudi Arabia is a cautious and conservative leader, prizing stability and the slow pace of events to rash actions. Its ample financial and oil reserves permit it. Saudi Arabia will behave as an influential member of OPEC, one of the leaders, but not the leader. It will, unless provoked beyond forbearance, exercise a calming and civilizing influence. Precisely because of Saudi Arabia, the world oil industry has not been and will probably not become a dominant-firm oligopoly.

OPEC Secretary General Ali M. Jaidah of Qatar told me in 1978 (Interview, *Worldview*, March 1979, p. 41) that "OPEC is not illegal, nor is it a cartel or an oligopoly. OPEC is OPEC. Something new under the sun." He is right. OPEC resists the usual forms of economic and political analysis. The search for an appropriate historical and theoretical model has come to naught. Predicting its behavior and future on the basis of existing models will be right only by accident. OPEC may also be immune to policies by oil-importing countries based on analysis that does not recognize its historical roots, complex economic, political, and psychological modus operandi, and diverse and multiple interests and goals.

MOVING DOWNSTREAM

The OPEC takeover sundered the chain that linked the well head to the gas pump. The break in the chain came at the water's edge of the exporting country. From that point back to the oil in the ground—upstream, as the oil men call it—belongs now to OPEC. From that point forward—downstream—through the refinery to the customer, including transportation and storage, the companies continue to exercise full sway. Some downstream operations, notably refining, occurs in the upstream area. And the companies also continue to operate upstream, under OPEC supervision, in member countries, and to control upstream operations in non-OPEC areas. The upstream-downstream division is not a sharp dividing line.

OPEC wants to move downstream more. In the development of the members' economies, refining, transportation, petrochemicals, and marketing can form the solid industrial base from which modernization and industrialization can spring. Members see the values added by manufacturing and services going to the economies of the industrial countries by refining OPEC oil, for example, and want to capture more of the value for themselves.

Members reason that if they cannot diversify their economies by the further processing of their own raw materials, then they have little hope of developing other industries. They point out historically that many of the present developed countries, the United States and Canada, for example, got their start by first exporting raw materials. Their development then proceeded to manufacturing in which their economies processed raw materials into finished goods for exports. OPEC members claim the right to develop along the same path.

In 1939 almost 70 percent of the crude oil produced in the world was refined at the source. During the second world war and in the ensuing great industrial drive in Europe, North America, and Japan, refining facilities moved to the market. New refineries were built at the market or at intermediate locations. In 1951 only 50 percent of refining capacity was at the source. By 1973 only 9 percent of world refining capacity outside North America and the Soviet-bloc area was located at the source. About 3 percent was at intermediate locations. The oil-consuming countries hogged 78 percent.

OPEC members intend to reverse this trend. From 1973 through 1977 they expanded their refinery capacity 14 percent. They have plans for further expansion, at an even faster pace than in recent years. Every OPEC member is currently building new capacity and expanding existing refineries. At present, OPEC owns or controls only about 6 percent of world refinery capacity, refining only about 11 percent of their own crude oil. Much of their current production of oil products goes for consumption within OPEC members, which is growing rapidly. Members want even greater expansion so that they can begin large-scale export of oil products.

The world refinery industry stands in its way. In oil-importing countries some excess refinery capacity now exists. In Europe, for example, refineries in 1977 operated at less than two-thirds of capacity. In Japan only four-fifths of capacity was used. The United States came closest to using its facilities fully, but even it could have expanded nearly 15 percent without adding more refineries. By early 1979 excess refinery capacity was down, especially in the United States, and under pressure of strong demand, shortages arose because of refining limitations. Still, given the present size and capacity of the refining industry in the industrial world, how can OPEC get a start?

The OPEC answer is simple. Europe, Japan, and the United States should close down their least efficient refineries and not build more. The expansion of world refinery capacity would then occur, with the help of the industrial countries, within OPEC. Industrial countries do not view this idea sympathetically. Indeed, they scoff at it. Refining is an important industry in the industrial countries and they have no intention of abandoning it to OPEC members. They may shut down some capacity. Most of it, however, they plan to absorb as consumption rises. They will expand capacity at home as the need arises. Industrial countries argue that the location advantages—costs of shipping crude to refineries and products from refineries to the market—favor the market location. Finally, they insist that the market location will be more efficient and more responsive to the needs of consumers.

Undaunted, OPEC counters every argument. In the interest of exporting more products, members are beginning to flex their muscles. Secretary General Ali M. Jaidah said in October 1978, that OPEC members were considering coupling the export of crude oil to the purchase of products also. *(Washington Star,* October 10, 1978, p. A-8). If importers do not buy the appropriate amount of products, then members will not supply the crude. It would be a move akin to that of Saudi Arabia in 1978 that required the purchase of heavy oil along with the light oil that is in greater demand. OPEC is only beginning to press its case for more refinery production.

Other downstream operations also interest OPEC. Private oil companies, through contracts and local service companies, still do four-fifths of the marketing of OPEC crude in world markets. Many of the companies are buying the crude oil for their own refineries at market locations. OPEC national oil companies market directly to governments and refineries abroad only about 18 percent of world crude oil. OPEC members want a larger share.

The Persian Gulf national oil companies are the most aggressive direct sellers, peddling more than 30 percent of their own crude abroad. These and other OPEC national oil companies still have, however, only limited downstream contacts and arrangements. It takes time to develop these markets. Refineries that are captives of the large companies that are often preferred OPEC buyers also limit the downstream movement of OPEC national companies. Still, whenever the opportunity arises, members export directly and insist on enlarging their share of world crude-oil marketing.

The OPEC national oil companies and their subsidiaries own and operate about 100 oil tankers with 13 million dead weight tons. This is less than 4 percent of world tanker capacity. The OPEC tanker fleet has expanded modestly in the 1970s but has rough going in the face of the large excess world tanker capacity. Indeed, the industry every year has been junking more tonnage than OPEC has. But some members continue to expand their small fleets and they are determined to increase their share of crude-oil transportation. If necessary, they may link crude sales to shipment in OPEC bottoms.

OPEC also wants a slice of the world petrochemical industry. At this time, it has almost none of it. Venezuela has two small plants. Iran in a joint venture with Japan has a partly completed plant. There are other small facilities and some more plants are in the planning stages. The petrochemical industry is a high-technology industry and requires the assistance of the advanced countries. Oil experts state that oil-producing countries have no advantage and, from the technological and cost points of view, many disadvantages in the production of petrochemicals. Still,

OPEC members want to produce some specialized products that they believe are within their capabilities and to wring whatever end products they can out of their crude oil.

In a world in which twenty countries possess, in addition to the market for most oil products, nearly all the technological and economic advantages, the efforts of OPEC to move downstream appear quixotic. This does not deter the OPEC members, who are thinking in long-run terms. The world, they say, must make way for them if they want to continue to use OPEC oil. Most experts expect OPEC to move downstream, but slowly, loosening only slightly the stranglehold that the industrial countries now have on the world oil-products industry.

OPEC members have in recent years accumulated great financial wealth. Members in that short time have been unable to translate the wealth into economic development. What has OPEC done with all its money, and how much is there, anyway? The next chapter examines OPEC's income and where the members have put their wealth.

Chapter 7

OPEC and Its Money

For a time in late 1973 and 1974, it seemed that the whole world had gone mad. The Arab oil embargo stirred popular fury in industrial countries and frightened the people. The press whipped it up, exaggerating shortages and repeating fearful doomsday predictions. The OPEC oil-price increase, an equally devastating though less publicized blow, stuck the banks and financial institutions. Billions of dollars flowed into the coffers of small countries not even used to handling millions, sucking dry the treasuries of other countries, and forcing half a hundred countries to borrow. Bankers trembled. Finance ministers panicked. World finance seemed to teeter on the edge of the abyss of monetary chaos.

The Shah of Iran committed $4 billion for the purchase of arms. Shiekh Mohammed al-Fassi, a twenty-three-year-old Saudi Arabian student at UCLA, bought a $2.4 million house in Beverly Hills and redecorated it garishly by community standards for another $1.5 million, scandalizing his staid neighbors. Kuwait bought some islands off the Carolinas. Saudi Arabia bought American businesses, farms, and U.S. Treasury notes. A sheikh bought a Rolls Royce for a tour of the British Isles. As he boarded his plane to return home, the driver asked him what to do with the car. The reply: "You can have it; it's yours." Bankers crowded into the twenty-fifth floor conference room of the Financial Tower in Caracas, eagerly trying to lend money. Newspaper columns filled with the quaint antics of the nouveau riche.

The financial experts wore long faces. They feared that by spending their money the new rich nations would wreck the world economy. They feared just as much that the new rich nations would not spend their money and wreck the world economy. The world financial community, already troubled with dollar sickness and the collapse of the Bretton Woods agreement that had guided world money markets for nearly three decades, had never before received such a shattering blow.

Many of the experts feared that the dollar would succumb and the system would break down, triggering both worldwide depression and wild inflation. Bankers worked frantically, kept mum, and hoped for a miracle. Even the most optimistic foresaw that Saudi Arabia and other OPEC countries would accumulate half a trillion dollars or more in surplus in a few years, more than enough to dominate world trade and finance. Someone coined the word "petrodollars." Novelists began cranking out end-of-the-world oil novels.

Few predicted what actually happened. In an orderly transition, world money managers and OPEC members cooperated in minifying the disturbances. They not only salvaged the financial system, but also proved its strength. Through it all, OPEC managed its windfall with aplomb. International finance grunted and groaned under the strain of transferring such vast sums and making funds available to those in need. In 1974 banks and other financial institutions moved the billions in payments for oil, enhanced OPEC imports, foreign aid, foreign investment, and swollen international reserves not only without cracking but indeed, with increasing confidence.

OPEC members turned the money around, giving a new meaning to the word recycling. Their imports increased rapidly. They deposited their money in experienced hands, created new institutions of their own, invested their revenues abroad, and mounted a foreign aid program. OPEC international reserves shot up, but not by enough to endanger the system. After 1974 the annual OPEC surplus not only did not increase, but rather began to decline.

The feared inexperience and irresponsibility of nations unused to high finance did not impair the operations of world money. There was no bungling. No threats to use oil money for blackmail or retaliation marred the transactions. What could have ended disastrously for both OPEC and the industrial countries in fact became a triumph for the world's bankers and finance ministers and the monetary system. By 1977 the experts had shifted their worries from OPEC money back to the shrinking dollar once again. A residual of the price increase of oil, however, continues to haunt the financial community. More countries are deeper in debt with less prospect for repaying than ever before. But the system survived.

EARNINGS FROM OIL

OPEC members earned nearly $700 billion from 1973 through 1978. This figure does not include their earnings from investments abroad, which are still relatively small. Export earnings of OPEC show no sign of either very large increases or of systematic decline. In 1978 earnings were down 3 percent from their 1977 all time high of slightly less than $150 billion, but were 3.8 times those of 1973. The largest increase in earnings came in 1974 when they jumped from about $40 billion to $120 billion. In the intervening years earnings went up and down. In the slump year of 1975 earnings fell by 8 percent, but the next year they rose by 12 percent. In 1977 they rose again by about 7 percent. About three-fourths of the increases in earnings came from price increases. At the 1973 price OPEC members would have earned about $200 billion.

Saudi Arabia and Iran have fared the best. The largest oil producer earned $181 billion from 1973 through 1978, the second largest $118 billion. The two combined account for 43 percent of all OPEC earnings. Even the smallest producers, Gabon and Ecuador, have each been earning more than $1 billion a year, five times what they earned in the early 1970s. In 1978 and 1979 most OPEC members were earning around $10 billion a year, putting them in the class of middle-sized world traders. Saudi Arabia with $41 billion and Iran with more than $24 billion in 1978 are in the big league among world exporters.

Before becoming too awed by these large figures, bear in mind that all OPEC export earnings are only 12 percent of the world's annual exports of almost $1.2 trillion. The increase in OPEC earnings—a little over $100 billion from 1973 through 1978—is only about 15 percent of the increase in total world exports. The United States spends more on defense each year than OPEC earns. Americans also spend more on welfare than the more than 300 million people of OPEC members earn. With all their wealth, the per capita incomes of OPEC members are still less than one-tenth that of the United States.

Critical dependence on oil exports marks the foreign trade of

OPEC members. Some members export small amounts of other goods. Oil is only about 70 percent of Indonesia's exports, for example, 80 percent of Gabon's exports, and only 50 percent of Ecuador's exports. For most members and for all the large producers, exports are more than 90 percent oil. Saudi Arabian oil exports, for example, are 99.9 percent of its total exports. For all OPEC members together, oil averages 94 percent of total exports.

The oil export earnings, paid in dollars, move into the bank accounts of national oil companies and other special accounts of OPEC members, or are paid by the companies directly into the national treasuries. Funds not already in the treasuries' hands move into the treasuries before disbursement. The finance ministries distribute the money to other ministries and to other governmental agencies responsible for authorizing the paying for imports and investments abroad.

The oil earnings underlie the domestic money supply of OPEC members, but the connection is loose. Most of the dollars appear as a credit entry to an account in a foreign bank, only to disappear days, weeks, or months later to pay for imports or to buy an asset. The central banks of OPEC members must increase the domestic money supply for those who trade rials, riyals, or bolivars for dollars. Government agencies, however, often simply receive their budget, or a part of it, in dollars in a foreign bank, circumventing the exchange of domestic for foreign currency. OPEC domestic monies, circulating through the economies of members, sustain private domestic production and consumption and encourage economic growth. Backed by dollars and other foreign currencies, the domestic currencies of some OPEC members have themselves become strong international currencies.

Export earnings are the lion's share of OPEC members' economies. Total Saudi Arabian production is only one-fifth greater than its oil production. Iran is more diversified. Its oil exports were only two-fifths of what the economy produced before they halted and resumed again in 1979 at a slower pace. Venezuela and Indonesia also have relatively complex economies. Even in a case such as Venezuela, where oil is one-third of what the economy produces, it is still the lifeblood that nourishes the entire economy. If by some stroke of fate the oil of OPEC members suddenly disappeared, those thirteen countries would be poor, powerless, and unpromising.

One element of exchange earnings that has increased in importance in recent years for some OPEC members is investment income. The large foreign holdings of Saudi Arabia, Kuwait, Iran, the United Arab Emirates, and some other members before and following the

1973 price jump have begun to pay off. In 1976 these earnings were about $7 billion, rising to over $8 billion in 1977 and $12 billion in 1978. These earnings reflect the financial assets bought with the surpluses of OPEC members, invested primarily in Western Europe and the United States.

IMPORTS AND TRADE BALANCE

The panic that ran through the world financial community originated in the uncertainty about what OPEC members would do with all their new wealth. If they did not spend it, a huge empty hole would appear in international finance. Funds would be isolated in negotiable instruments and in financial institutions, or even in hoards of gold or other materials, not necessarily returning to the flow of international finance. That flow is necessary to sustain the financing of world trade. An interruption would strain both lenders and borrowers, disrupting imports and exports.

If, on the other hand, OPEC members spent their money quickly enough to be available to others, financiers worried about whether or not the funds would pass through channels and come to rest in a way that would facilitate their use in trade. Many countries faced much higher oil bills and had to borrow. Unless the OPEC money came back into circulation through the normal channels of world finance, financial institutions would not have the support for funds to lend to those needing money to pay their bills. The swiftness of the change to debtor positions for oil importers and strong creditor positions for OPEC members startled the international bankers. They were not prepared for the change in direction of this magnitude of money flows.

An almost immediate surge of imports by OPEC members, however, replaced some of the drain of money used to pay for the oil. OPEC imports in 1974 increased nearly 70 percent. This was not nearly as much as the tripling of exports, but it still represented a large increase and foreshadowed how OPEC members would use their new resources. In 1975, the slump year of OPEC exports, members increased imports by nearly 60 percent more. And as each year passed, they went up more. In 1977 OPEC imports were $83 billion, and in 1978 they were $101 billion.

For the period from 1973 through 1978, imports by OPEC members, largely from the industrial countries, were $360 billion, more than one-half of the members' export earnings. Iran, with its more diversified and advanced economy, led the way in imports.

That country imported more than $64 billion in goods. Not far behind was Saudi Arabia with nearly $62 billion, with a great surge of imports in 1978 of 52 percent. Venezuela and Nigeria followed with imports of about $42 billion each. These four countries accounted for nearly three-fifths of all OPEC imports.

Bear in mind that these are imports only of goods. OPEC members also must import substantial amounts of services. Some of these service imports are tied to the amount of exports. They are payments to oil companies for transportation, marketing, technology, technical assistance, and financial and other services. Without the services represented by these import payments, the oil exports would not be possible.

Some service imports are money paid to foreign firms for technical services in support of development programs. The payments for more than 50,000 students from OPEC members studying abroad, tourism, and other services are also in this category. The tour by the sheikh of the British Isles, along with the Rolls Royce he gave to his driver, was a service import. Nearly one-fourth of total exports go to pay for service imports. In 1976 the bill was $31 billion; in 1977 it was about $38 billion; and in 1978 it was $43 billion. OPEC members paid out from 1973 through 1978 more than $170 billion for service imports, nearly one-half the amount they paid for goods imports.

That makes the full import bill of OPEC members from 1973 though 1978 over $530 billion. The difference between exchange earnings and payments for imports was about $170 billion, less than one-fourth of exports. Put the other way around, OPEC members have spent about 76 percent of their new money on imports and have saved and invested the rest. The gap between imports and exports has steadily closed. By 1978 most OPEC members had once again become debtor countries, and the few not borrowing had only a tiny surplus. In 1979, however, the surplus will grow once again. Exports earnings will climb to at least $175 billion, largely because of the price increases. Although imports will also increase, the surplus for the year will be $30 billion or more.

Imports went to increase consumption, to add to domestic assets through additional capital and intermediate goods, and to increase the supply of arms. Little is known of the arms bill, but it runs into billions. Most of the funds, however, were used to expand members' development and industrialization programs. For example, fixed capital formation in Saudi Arabia between 1973 and 1977 increased more than tenfold, while private consumption increased only 3.2 times. In Iran in the same period, private consumption went up less than one-half as fast as capital formation. Although OPEC members indulged in some wasteful and conspicuous consumption, the largest slice of their new wealth has gone to promote economic development.

The surplus in the current international accounts of OPEC has declined and almost vanished. In 1974, the year of the largest price increase, OPEC members spent $64 billion less than they earned. Some feared that the large surplus would continue and that in a decade OPEC would have an unbelievable hoard. But the next year, the surplus plummeted to $31 billion as a result of rising imports and falling exports. In 1976 the surplus edged up to $35 billion, reflecting a modest recovery, but in 1977 it was back down to $30 billion. In 1978, reflecting continued rising imports with exports in the doldrums, the surplus was only about $10 billion. At least in the near future, few expect OPEC to accumulate large surpluses.

The boom year of 1974 gave all OPEC members a surplus. But the next year, two countries—Libya and Algeria—were already running a deficit in international accounts. Algeria has continued in deficit. In 1976 several OPEC members, including Nigeria, Gabon, Indonesia, Venezuela, and Ecuador, began to run deficits. The deficit of this group of countries has increased each year. The strongest surplus country is Saudi Arabia, with one-third to one-half of the total OPEC surplus each year. Kuwait and the United Arab Emirates also have strong surpluses, as have Iraq and Iran. The 1979 surplus will be about the same as the 1975-1977 surpluses. But the prospects for continuing large annual surpluses are not great.

One of the more troublesome problems that confronted international finance was the mismatch of OPEC imports and exports. OPEC members often do not sell their oil to the same countries from whom they buy. Italy, with a trade deficit of $5 billion in 1973, had a deficit twice as large in 1974, attributable largely to oil. Yet Italy exports little to OPEC members. On the other hand, Germany had a surplus of $13 billion in 1973, and in 1974 its surplus was nearly $20 billion. Germany exports a lot to OPEC members. Many countries with balanced imports and exports before the oil-price increase suddenly had deficits and those already with deficits often had larger deficits.

The less-developed countries faced the hardest adjustment. Their deficit of $12 billion, about 18 percent of their exports in 1973, shot up to $33 billion in 1974, 33 percent of exports. But despite the recession and inflation in advanced countries, they managed to increase exports and hold down imports. In 1977 their deficit was down to $26 billion. In 1978 it was back up to over $40 billion. Only the price increases of many primary products exported by less-developed countries at the same time as the oil-price increase, as well as borrowings, saved many poor countries from economic disaster.

It has taken time for a new normal pattern of trade and finance to emerge. The process is still not complete, but the maladjustments are no longer as threatening as they were in the recession years of 1975 and 1976. Many countries whose imports from OPEC have been large have been able to increase their exports either to OPEC, or they have exported more to countries that do export more to OPEC. Other countries have sternly reduced their imports from OPEC while exporting more to members. Most trading nations have participated in the large general expansion of international trade in the years from 1973 through 1978—an increase of 127 percent.

Even so, some countries, especially the less-developed countries, go deeper and deeper into debt. Others, such as Germany and Japan, frantically push exports to maintain their position. Nearly all of the industrial countries have recovered from the 1973-1974 price increases, but all have smaller surpluses or larger deficits than they had a few years ago. The United States, alone among the great powers, goes against this trend. It has had increasing deficits. The U.S. trade deficit was only a little more than $2 billion in 1973, registered a surplus of nearly $4 billion in the recession year of 1975, but jumped to a deficit of $36.5 billion in 1977 and $39.4 billion in 1978. Since 1973 the total debt of the world has substantially increased, posing the threat of financial instability to some countries and imposing hardships on others.

FOREIGN ASSETS

The foreign assets of OPEC members at the end of 1978 stood at about $180 billion. One of their principal assets is their holdings of international reserves. These are funds in bank deposits and short-term financial instruments in foreign currencies, mainly the dollar, the yen, and European currencies. The purpose of international reserves is to cushion the domestic economy of a country against changes in imports and exports and to defend the country's exchange rate. Although the need for reserves has diminished now that countries permit their exchange rate to vary and the value of their currencies to change according to supply and demand, most countries still hold reserves. They often earn less than long-term financial assets and real assets, but they facilitate trade and payments and are a guarantee against contingencies.

OPEC international reserves at the end of 1978 were over $60 billion, down from an earlier high of $75 billion. OPEC's holdings are greater than those of Germany and more than those of the United States and Japan combined. Saudi Arabian reserves are only slightly less than those of the United States. OPEC reserves, one-third of their total foreign assets, are a much higher ratio of reserves to foreign assets than the industrial countries maintain. The high OPEC reserves reflect the short time over which they were earned and members' inexperience as foreign investors.

Saudi Arabia with $20 billion in reserves has more than Iran with $12 billion and Venezuela with $7 billion combined. Some countries hold international reserves even though they have borrowed. Algeria, Indonesia, Gabon, and Ecuador, for example, have reserves although they have used up their surpluses of earlier years. Iraq with $6 billion, Libya with $4 billion, and Nigeria with $2 billion probably have more of these liquid assets than is justified by their small accumulated surpluses. They, like the other members, maintain reserves to assure liquidity.

In addition to these reserves, OPEC members also own funds on deposit with, or have lent money to, the International Monetary Fund (IMF). Their quotas, recently raised to reflect their new more prosperous status, total nearly $4 billion. OPEC contributions to the IMF special oil and supplementary facilities are nearly $6.2 billion, of which $4.6 had been used by early 1979. In case of need, OPEC members may borrow against these funds, in addition to their normal borrowing privileges. OPEC members also lend to the World Bank and other international financial institutions. OPEC lending to IMF special facilities, purchase of World Bank bonds, and pledges to other financial institutions come to more than $15 billion.

The amount of total foreign assets owned by OPEC members matches approximately the surpluses that they have earned since 1973. Some differences arise because of surpluses and deficits prior to that year. Kuwait, for example, whose foreign assets exceed its recent accumulated surpluses, had large holdings carried over from before 1973. Nearly 90 percent of OPEC foreign assets consist of its international reserves, as well as identifiable bank deposits, portfolio investments, and ownership of financial instruments in the United States and other large industrial countries.

More than one-half of the identifiable assets are in bank deposits.

These are in dollars, pounds, Canadian dollars, Japanese yen, other European currencies, and in the European currency market. The dollar predominates. More than $25 billion is spread out in American treasury bills, notes and bonds of the United States, Great Britain, Japan, and Canada. Ownership of property and enterprises in the United States and other industrial countries, as well as direct loans to governments and enterprises account for only a little more than $20 billion.

The foreign assets of OPEC members may appear to be large, but the wealth embodied in the international financial community dwarfs them. In early 1979 OPEC members held only 16 percent of the total international reserves of $370 billion, and their total export earnings for 1978 were only 12 percent of world exports. The money and bond markets of the United States, Great Britain, Germany, Japan, and several other countries are many multiples of the OPEC funds they have absorbed. The ownership of property and real assets by OPEC members in the industrial countries is miniscule. Nowhere in world financial markets do OPEC members dominate.

The oil-exporting countries have behaved as prudent, even conservative investors. Most of their investments are in the form of relatively low-yield assets that are quickly available as money. Their ownership of long-term assets—property, direct loans, equity—that offer higher yields is quite modest. Their confidence in the industrial countries is shown by their holdings of government bonds. Their responsibility to the international community is shown by lending to international organizations. If their foreign investments display anything unusual, it is the relatively high concentration of assets in bank deposits, a characteristic of new savers.

High revenues have provided OPEC members with the opportunity to fortify their participation in international organizations. Their quotas—shares—in the World Bank have gone up and they have lent the bank money. Venezuela, for example, has lent the World Bank more money than its previous loans from the bank. The quotas in the IMF have also increased and two OPEC members—Saudi Arabia and Iran—have currencies that now count in the basket of currencies that constitute the Special Drawing Right (SDR), the IMF unit of account. OPEC members were instrumental, through their contribution of $436 million, in creating the International Fund for Agricultural Development.

Morgan Guaranty Trust, whom the OPEC secretariat regards as reliable, estimates that Saudi Arabian foreign asset holdings at the beginning of 1978 were $68 billion, 40 percent of the total OPEC foreign assets. The other three Persian Gulf Arab countries—Kuwait, the United Arab Emirates, and Qatar—also own large foreign assets. The total of these four countries is 71 percent of the OPEC total. Add Iran's $22 billion and the five countries account for 84 percent of OPEC foreign holdings.

Four countries are in a net deficit position and others are moving in that direction. Algeria, Ecuador, Gabon, and Indonesia had a negative external asset balance in 1977 of $15 billion, $9 billion of which was attributable to Indonesia alone. The net asset position of OPEC in 1978 was about $160 billion. Growth in that year was slight. The price increases and strong demand of 1979 may result in an enlarged surplus, especially for the Persian Gulf countries. But it is difficult to imagine circumstances in which OPEC members' galloping surpluses and the accumulation of very large foreign assets could continue indefinitely.

IMPORTS AND DEVELOPMENT

OPEC members do not want to sell their oil just to have money. Oil exporters want to trade one asset—oil in the ground—for other assets such as buildings, machinery, factories, roads that increase the income and productivity of their economies. Whatever else OPEC members do with their money, the highest priority expenditure for all of them has been the purchase of imports for economic development. Even the OPEC members richest in money have underdeveloped economies. All wish to build their economies to the point where their agriculture, industry, and commerce can continue to grow when the inevitable day arrives when the oil is gone.

Economic development is not simple or easy. It requires resources such as land, water, and raw materials. It requires people—unskilled, skilled, professional—to do the work. It requires machinery and equipment, plants and power lines, homes and schools. At the very least, it requires the ability to put the resources, labor, and capital together to produce goods. In order to develop the economy, a country must have or acquire the technology to build and operate its industries, have an organization that facilitates production, and have the will to break the chains of underdevelopment.

Every OPEC member lacks one or more of these ingredients. Some have more than enough people but the workers do not possess the skills for modern production. Many have little arable land, not enough water, and few raw materials besides oil. All of them lack capital and most have insufficient enterpreneurial ability. Their own technology is primitive. Few have economic organizations and institutions adequate for modern production. The leadership of all have the will to develop, but they must continually fight lethargy, poverty, and tradition's dead hand among the people.

Economic development demands its ingredients in a special combination. It does no good to have sophisticated lathes and boring machines if there are no workers to operate them. It does no good to have canning factories if agriculture does not produce the food for them. It does no good to have electronics engineers if there are no plants for them to work in. And it does no good to have highways that go nowhere, with no trucks to run on them and no produce to haul. Economic development is a jigsaw puzzle, with all the pieces fitting together just so. The economies of OPEC members have many pieces missing and those they have do not match.

Imports play a vital role in economic development. They supply the missing pieces of the puzzle of development. They are the scissors that trim the pieces to fit properly. With enough resources in the form of money for imports and full use of their own real resources, the economies of OPEC members can develop. What is lacking at home comes from abroad. The process, however, is not instantaneous. It takes time to build roads and factories, to train people, and to organize production. Some OPEC members are better equipped to make use of imports in a short period than others. They can import and put the imports and domestic resources to work quickly. Other OPEC members will require a long time.

How long it takes depends upon how primitive the economy is, and on how many pieces of the puzzle are missing or don't fit. The rate at which imports can be put to use is called the capital absorptive capacity by economists. A country with a low capital absorptive capacity, such as Kuwait, can increase its productive imports only slowly, even though it may have unlimited amounts of money. A country with a high capital absorptive capacity, such as Nigeria, can increase imports rapidly. Of course, any country can import for consumption or arms purposes. Spending on imports to increase the economy's productive capacity, however, must conform to the measured pace at which imports can be assimilated.

Among OPEC members there is great variation in their ability to absorb and productively use imports. Algeria, Iran, Nigeria, Indonesia, and Venezuela all have high capital absorptive capacities. Saudi Arabia, Kuwait, the United Arab Emirates, Libya, and Qatar have low capital absorptive capacities. No exact correlation exists, but countries with large financial surplusses at present are the ones that tend to have low capacities and the countries with high capacities spend all they earn and even borrow.

OPEC PLANS AND
FINANCIAL NEEDS

The oil-price bonanza induced all OPEC members to raise their economic planning sights. The new revenues permitted them not only to undertake new projects, but also to revise their entire strategy and timetable for development. Most hurried to produce new five-year plans to put the oil income to work. The results are reflected in the greatly enlarged imports of OPEC members, in the bustling prosperity of their economies, and in their economic growth rates, which are the highest in the world.

Each member faces constraints in its ability to absorb the infusion of new money into its economy. No matter what the plans say, no matter how much money is available, only a certain amount can be spent by each country in its drive to development without waste. And in just a few years, a country can do little to alter its ability to employ new income productively. As the economy develops, its ability to undertake new development projects expands. The countries of OPEC differ greatly in their financial needs.

Saudi Arabia, for example, plans to spend $142 billion between 1975 and 1980. This amount is easily within the country's financial capacity. Indeed, some financial surpluses would continue to accumulate with spending at that rate. But some question exists with respect to Saudi Arabia's ability to spend even that much without breaking the bottlenecks in its economy.

One of the principal constraints on members' development, including that of Saudi Arabia, is the paucity of labor at every skill level. In order to accomplish its plan, the country must import labor in large amounts and make provision for their housing and social needs. These expenditures are not contemplated in present plans. Saudi Arabia will either spend less than $142 billion and grow less, or, if it resolves its labor shortages will spend even more in meeting its target growth.

Labor also constrains the Iranian economy but to a lesser degree. Still, its gross national product could begin to decline unless it imports large amounts of skilled labor and increases the productivity of its own labor force. Iran is a large country, populous yet poor, with a diversified economy so that its imports must be high and growing even to achieve modest growth. Nigeria will also be able to translate its oil revenues into relatively rapid growth, but only if it acquires sufficient skilled labor. Improvements in labor productivity are needed if the Indonesian and Iraqi economies are to grow. Ecuador needs to mobilize its unemployed labor force to achieve rapid growth.

The financial needs of some OPEC members have already outstripped their revenues. In the euphoria of the deluge of money, some countries have also wasted money on useless projects and permitted domestic consumption to rise rapidly. Some of these measures do indeed improve the welfare of the people momentarily, but do nothing to strengthen the economy or increase its prospects for development. Nonproductive expenditures of oil revenues continue to characterize the spending of OPEC members. In addition, the oil revenues of some countries have levelled off or, as in the case of Venezuela and Kuwait, declined.

The squeeze between expenditure and revenues has forced many members—Algeria, Indonesia, and Nigeria, for example—to borrow. The amounts, compared to revenues, are still small and are within the capabilities of these members to repay. In 1977 all OPEC members borrowed $11 billion, and in 1978 this figure was up to $18 billion. Venezuela, Algeria, Iran, and Nigeria led the borrowers. The borrowing has been in all cases for specific projects designed to increase the productivity of the economy. The lenders in most cases have been large international bankers. Without large infusions of new revenues, most members—all except Saudi Arabia and the Persian Gulf Arab countries—will be borrowing in the near future.

The rapidity with which many OPEC members have absorbed their surplusses and begun to borrow reflects their underestimation of the costs of development. Some of the development plans did not provide for all aspects of development, such as social and administrative services. Most plans did not make enough allowance for inflation in the prices of imports. The impact of government expenditures on the private domestic sector was overestimated. Venezuela's five-year plan (1976-1980) contemplated more than 8 percent growth from $53 billion in government investment and about the same in private investment. Already it is evident that private investment will lag seriously, resulting in not more than one-half of the projected growth.

The countries with large reserves, such as Saudi Arabia, Kuwait, Qatar, and the United Arab Emirates, have borrowed only trivial amounts. Still, the inflated costs and underestimates of development needs have eroded their surpluses. For either the surplus countries or those that are borrowing, the availability of finance has not restrained their growth. In 1978, however, bankers began to have second thoughts about lending to some members and became more conservative. Physical constraints, such as lack of land, water, labor, low labor productivity, and other elements not remedial in the short-run by imports, hold back greater economic expansion.

CURRENCY DEPRECIATION
AND INFLATION

The oil companies and OPEC have always priced oil and received payment in dollars. Five of the large oil companies are American, and the United States is the largest consumer of oil and the largest oil-importer, as well as one of the largest producers. The overwhelming importance of the American economy and the traditional stability of the dollar has made it the natural oil currency. As development needs press on revenues, this may change.

When the dollar was in trouble in 1971-1973 and the United States devalued the dollar twice, the oil industry clung to the dollar as the means of payment. In that period, however, OPEC and the companies devised a formula for revising the dollar price automatically as its value changed. When OPEC won the price battle in 1973, the unmodified dollar returned. Since then OPEC members have muttered and complained about the deterioration in the value of the dollar, but have done nothing.

The dollar satisfied OPEC members so long as its value, relative to other currencies, remained stable. They could use their dollars to buy the currencies of the countries from which they wish to buy imports or investments. But as the dollar's value sagged in 1977 and later, they had to use more of their dollars to buy other currencies, in effect raising the prices of German, French, Italian, and Japanese imports and investments. Depreciation also reduces the value of their dollar investments and their earnings.

The dollar in fact has lost ground compared to the yen and most European currencies. Unrestrained imports of oil, occasioned by low prices of domestic oil, as well as high imports, low exports, modest productivity increases, and the massive accumulation of dollars in the international economy have so weakened the dollar that OPEC members have begun to count their losses in sticking to the dollar in the billions of dollars. Throughout 1977 and 1978 officials of OPEC members commented that unless the United States did something to halt the dollar decline, OPEC would be forced to take action.

The natural solution for OPEC would be to establish prices in another currency or in a group of currencies. They would retain the dollar as the means of payment. Such a move would halt the erosion of investment values and higher prices of non-American imports. To do so could also trigger, however, further declines in the value of the dollar and threaten international financial stability. Since OPEC flourishes best with stable finance, it is a move that members do not relish.

The International Monetary Fund maintains a unit of account against which all currencies are measured. When initiated in the late 1960s, the purpose of Special Drawing Rights was to add to the funds available for borrowing by IMF members. Gradually, the SDR has emerged as a new international unit of account. It is composed of sixteen currencies and its value is calculated every day by the IMF. The dollar has the most weight, 33.3 percent, and West Germany, Japan, France, and Great Britain's currencies together account for another 35 percent. It is the fixed measure against which all currencies are compared. It is also a possible replacement for the dollar in the pricing of oil.

In November 1971 the dollar and the SDR had the same value. The decline of the value of the dollar with respect to other currencies had gradually altered the SDR value of the dollar. The value of the dollar declined to 86 percent of that of the SDR by the end of 1976. In early November 1978, the dollar's value was only 78 percent of that of the SDR. At the same time, the value of most other currencies climbed with respect to the SDR and the dollar. The German mark, Japanese yen, and several European currencies have experienced steadily increasing value relative to the SDR. The problem of the SDR for OPEC purposes, however, is that it reflects the general trading and financial position of countries, not the trading pattern of OPEC imports and exports.

Oil is becoming more important in trade and finance. Two OPEC currencies in April 1978, began to be counted among the currencies constituting the SDR. The Saudi Arabian riyal and the Iranian rial are still a small element in the SDR but the two are greater than some of the smaller but traditionally important European currencies. OPEC has considered establishing a basket of currencies of their own for pricing purposes, rather than using the SDR. But despite all the talk and substantial loses, OPEC is reluctant to cut loose from the dollar.

Secretary General Ali M. Jaidah explained it best. "The United States has lost some of its competitive advantage in competitive markets, partly through inflation and partly through failing productivity .... The result is deficits in the balance of trade and payments, falling demand for the dollar and strong demand for other currencies ... OPEC would abandon the dollar in favor of a basket of currencies with great reluctance. The U.S. is the greatest power on earth, and the dollar, despite these transitory difficulties, is a strong currency" (interview, *Worldview*, March, 1979, p. 43). He adds that unless the dollar stabilizes and recovers, OPEC will probably be forced to abandon the dollar.

The companion complaint of OPEC is the inflation in all developed countries. U.S. export prices climbed more than 60 percent from 1973 to early 1978; German prices climbed 65 percent, and Japanese climbed about 60 percent. British prices are up more than 85 percent. All manufacturing and capital goods prices have gone up much faster than has oil since 1974. In addition, OPEC claims that their import prices have gone up even faster than have those of other importing countries.

Mr. Adan al-Janabi in the April 1978, *OPEC Review* (pp. 8-14) claims that OPEC import prices have gone up astronomically. He says, "the price of exported oil has moved from 100 to 121, while the price index for the basket of goods and services on which the income from the sale of oil is spent has moved from 100 to over 300" (p. 10). Partly this is because the goods and services OPEC imports are different from those of industrial countries. But partly it reflects, he claims, discrimination against OPEC.

Regardless of whether or not industrial countries discriminate against OPEC, the combined effect of the decline in the value of the dollar and imported inflation has since 1974 substantially reduced the real value of a barrel of oil. OPEC members believe that this reflects the unwillingness of the United States and other industrial countries to live within their incomes and to resolve their internal economic problems without the unfavorable impact on world trade. The threat to international financial stability, OPEC argues, comes not from OPEC but rather from the oil-importing countries. The 1979 oil-price increases, however, have restored much of the loss, although the price still stands below the 1974 price in purchasing power.

OPEC AND FINANCIAL INSTITUTIONS

Despite the huge wealth available to OPEC members, their own banking institutions are anemic. Each country has a number of commercial banks and some have other types of financial institutions, many of them created in the last few years. The central banks and finance ministries of members have international banking connections and some countries permit foreign banks to operate in their territories. Domestic banks also have international correspondents but they do not handle the country's financial wealth. The revenues move directly into foreign assets in institutions in the United States, Great Britain, Japan, and European countries as assets of the treasuries and central banks of OPEC members. No bank in any OPEC member is among the important world financial institutions.

OPEC money is deposited in every major financial center in the world, usually in the larger banks. All OPEC members together put nearly 10 percent of their 1977 surplus into foreign currency deposits in Great Britain, but only a negligible amount was in sterling deposits. OPEC members placed 26 percent of their 1977 surplus in European and Japanese banks. These funds exist, of course, to facilitate payments for imports. They also reflect the degree of confidence OPEC members have in the various world currencies.

Wealthy individuals from some OPEC members have purchased all or part of some banks in industrial countries. The most celebrated case was the sale, by Bert Lance, former aide to President Carter, of his stock in the First Natonal Bank of Georgia to a citizen of Saudi Arabia. Although the transaction attracted great notoriety, it hardly indicates a move by OPEC members to acquire banks or financial institutions in Europe and the United States. The governments of OPEC members have shown no interest in using their money to control foreign financial institutions.

Members of OPEC are members of the World Bank, the International Monetary Fund, and most other international and regional financial institutions. Until 1974 their role was modest and inconspicuous. Since then they have used such institutions as an outlet for their funds and as a forum for their policies. In the last revision of the IMF's quotas the share of the thirteen OPEC members increased from 5 to 10 percent. Their combined participation of nearly $5 billion is now exceeded only by the United States. They have also contributed heavily to IMF efforts to lend to countries whose deficits originate in their need to import oil.

OPEC quotas in the World Bank are also going up and OPEC members have supported the bank with its financial resources. Arab OPEC members also participate in the Arab Social and Economic Development Fund, in the new Arab Monetary Fund (AMF), and in other regional organizations. The AMF is a regional copy of the IMF. Latin American OPEC members have increased their participation in the Inter-American Development Bank. The Opec Special Fund, an account for making loans to less-developed countries, uses most of the regional and international financial institutions to implement its program of general support and project loans.

Special investment funds and departments in central banks administer the oil revenues for OPEC members. Local commercial banks handle some of these funds on behalf of local government agencies. Nearly all of the OPEC money, however, finds its way into financial institutions over which the members have no control. As important clients, of course, OPEC members receive solicitous treatment. Indeed, bankers tend to make much of officials from OPEC members in an effort to obtain and retain accounts and to lend money. But OPEC's role has been limited to that of a customer.

POTENTIAL FOR MISCHIEF-MAKING

Imagine the chaos that OPEC money could create. Suppose that at a time when the dollar was weak and sliding down, the members of OPEC got together and sold their dollars, buying yen or marks or gold. The dollar would plunge, the other currencies and gold would shoot up, and international money markets would be in shambles. Suppose that OPEC decided to buy up all the gold in the world. The gold price would skyrocket, the value of even strong currencies would fall, and the disruption would imperil the international monetary system. Suppose that OPEC members switched their bank accounts into and out of various currencies on short notice to take advantage of small changes in currency values. They would create havoc in the banks.

All these and more are within the capability of OPEC. A concerted effort on the part of OPEC could probably bring down the international monetary mechanism. Here is the substance of a dozen potential disaster novels, in addition to those already on the paperback shelves. Although the world monetary system demonstrated its toughness in adjusting to the changes of 1973 and 1974, it still depends on the rational behavior of nations and institutions. Just as a terrorist with a hand grenade can destroy a huge jet, so OPEC could undo the sophisticated method for managing world money.

But OPEC, like the terrorist, would destroy itself in the explosion. The value of OPEC money and their investments, the development of OPEC economies, and the welfare of the peoples of OPEC members depend on the smooth functioning of the international economy. If its monetary mechanism were reduced to rubble, world trade, including exports of OPEC and the imports on which its members depend to develop, would be disrupted. An OPEC monetary assault on the established means of world finance would be suicidal.

The record of rationality and the support and use of the traditional means of world finance implies that such an assault by OPEC would not be credible. In addition, the survival capabilities of the large industrial nations with greater productive capacity means that

OPEC would suffer the most. As OPEC becomes more and more enmeshed in the world economy and as its funds are used in domestic development and foreign investment, the prospect for misbehavior by OPEC diminishes. Although OPEC will use its resources to pursue its goals, these goals are not only consistent with but also require stability in the world economy. OPEC is now part of the establishment.

So long as OPEC retains its integrity, its financial prospects appear good. Modest price increases, tagging along after the rate of inflation in industrial countries will provide the oil exporters with sufficient income to insure rapid economic growth for the next few years. Some members will borrow and others will lend in the international economy. These financial flows will not endanger the economies of industrial countries nor put international finance at risk.

OIL PRICES, INFLATION, RECESSION, AND GROWTH

Most of the ills that have befallen the economies of the United States and other industrial economies are often laid at OPEC's door. Inflation, recession, and slow growth have plagued those economies since 1973. OPEC makes a nice scapegoat. According to the analysis of many economists and politicians in industrial countries, the oil-price increase of 1973-1974 caused inflation which culminated in the recession of 1974-1975 and the reduced economic growth since. That these events have happened is not at issue. But did OPEC do it?

The economies of the United States and other industrial countries badly malfunctioned after the 1973 price increase. Production declined. Unemployment rose. Consumer prices shot up. The United States, as well as the rest of the world, suffered both inflation and recession. The United States did not recover until 1976. But with recovery came slow growth and continued unemployment. None of the world economies have regained the head of steam they had before the oil-price increase.

Consumer prices in the United States increased 11.4 percent in 1974 and another 9.2 percent in 1975, having increased in 1973 only 6.2 percent. Gross national product declined, in constant prices, 1.7 percent in 1974 and another 1.8 percent in 1975, having increased in 1973 5.4 percent. Unemployment, which averaged 4.8 percent in 1973, increased steadily in 1974 until mid-1975 when it reached 8.7 percent. Investment declined from a high at the end of 1973 by 17.3 percent in late 1975. Consumer spending declined only 1.2 percent, but durable goods spending by consumers declined 12.7 percent between the end of 1973 and the end of 1974. The federal deficit, $14.4 billion in 1973, declined to $3.5 billion in 1974, but increased dramatically to $44.2 billion in 1974.

The crude-oil price increase was a part of the inflation, not its cause. When the crude-oil price quadrupled, true, the prices of products using imported crude oil also went up, as well as the prices of products using products made from crude oil. The effect rumbled through the economy but the indirect price increases dwindled quickly. It is easy to exaggerate the effect of the initial price increase. The domestic crude-oil price, fixed by law, dampened the effect since domestic crude was two-thirds of U.S. consumption. The gasoline and motor oil price index went up about 25 percent between the third quarter of 1973 and the end of 1974. The corresponding increase for fuel oil was only 45 percent. Domestic oil production, unchanging tax rates, and more stable other costs sheltered the American economy from the effects of the price increase.

Various estimates have been made of the price effect. Most indicate about 2.5 percent in 1974 and 0.5 percent in 1975. Since prices went up 11.4 percent in 1974, the calculations showed that prices would have gone up about 8.9 percent without the crude price increase, a jump in the inflation rate over 1973 of 44 percent. The price increase for 1974 and 1975, 21.6 percent, would have been 18.6 percent without the oil price increase, compared to 13.9 percent in 1972 and 1973. The boost in inflation in many other countries, not shielded by price stability in domestic production, was considerably greater than in the United States.

Inflation arises when people—consumers, investors, exporters, and governments—try to spend on available goods and services more than their value at existing prices, and either have or borrow the money to effect the transactions. Competition for limited goods forces prices up. When the price of one good increases many things besides inflation can happen. Indeed, when one price—the price of government, taxes—increases, most economists argue that inflation will slow down. The OPEC price increase of 1973-1974 can be viewed as a tax—an international sales tax on oil.

If people had bought less imported oil, or, having bought the same amount of imported oil, bought less of other goods, inflation need not have resulted. If total expenditures had remained the same, lower sales and falling prices would have offset the increasing price of oil. The reason prices went up after the oil-price increase was that people tried to increase total spending without production to match.

Total spending can exceed total production and create inflation if the spendable income generated by production exceeds the value of production of final goods. Some production, such as those for armaments, do not result in goods on the market. Taxes paid from savings and payments to welfare recipients also creates income unmatched by production. Wage increases in excess of increases in productivity means that wage earners can spend more than they produce. Borrowing against future income also generates an inflationary potential.

All of the industrial countries were caught up in an inflationary spiral long before the OPEC price rise. The inflation was fed by wage and income increases not matched by increases in production. The oil-price increase was only one event in the on-going inflation. It need not have added to inflation. The degree to which it did add to inflation reflects not the price increase, but rather the efforts of people and nations to continue to spend more than they produce.

How can people spend more when they do not produce more? Easy—they borrow. But it must be a special kind of borrowing. It must add to spending without adding its equivalent in production. If the lender foregoes spending in order to lend or if the borrower uses the loan to produce more, then spending and production can still match, even with borrowing, but with no inflation.

The typical method for creating purchasing power without production is by creating money. The new money permits additional spending without additional production. This is a part of what happened when OPEC increased the price of oil. Countries went on spending what they had spent before as well as spending to pay for the oil at the new price. The purchasing power that the industrial countries transferred to members of OPEC did not significantly diminish the spending by industrial countries. They borrowed and created money. OPEC created the possibility that the industrial countries would respond by inflation. But OPEC did not create the inflation. The industrial countries did that.

In response to the price increase, people also switched expenditures. They spent more on oil products and less on other goods. Purchases of many consumer durables and of residential construction, both of which played a prominent role in the recession as well, declined. But strong unions and oligopolistic enterprises not only kept prices in these sectors from falling, but rather increased them. The decline in production brought no corresponding reduction in spending.

Until recent decades, inflation and recession were opposites. Inflation takes over in a prosperous fully employed economy that is growing rapidly. In a recession, with falling spending, unemployment, and stagnation, inflation was supposed to halt. But industrial countries now have recession and inflation at the same time, as they did in 1974-1975. Spending was so high that inflation continued and even accelerated, but was not high enough to keep the economy fully employed and prosperous.

Investment in the United States was in trouble before the oil price increase. Throughout 1973 investment spending stagnated. Residential construction, an important part of investment, began to fall off. Inventories, on the other hand, climbed, indicating that final sales were not being realized at expected levels. These are signals for a retrenchment in investment spending. Even if the problems associated with the price of oil has not occurred, the economy would have declined in 1974. The recession was made in the United States.

With sufficiently stern measures, the industrial countries could have restrained inflation and avoided the worst effects of the recession. Less borrowing to pay for oil imports and, as investment slumped, more borrowing to stimulate production would have helped. Limitations on oil imports, price controls, and rationing would have kept inflation down. These kinds of policies are, however, extraordinarily difficult to initiate and implement. Still, the failure of the industrial countries to respond appropriately makes them just as responsible for the inflation and the recession as the members of OPEC.

The effects of inflation and recession in industrial countries spread throughout the world. The less-developed countries, already in economic torment, suffered a critical setback. Already they felt that the industrial countries were taking advantage of them. The fallout of the OPEC price increase included a resurgence of the desire of the poor countries to reform the international economy. This is the topic of the next chapter.

Chapter 8

OPEC and the Poor Countries

The impersonality of the market place is both a virtue and a vice. The market cares not whether the buyer is rich or poor, so long as the price is paid. Rockefeller may shrug off the price as trivial. To the ghetto black the same price may spell economic disaster. Two prices, one for the rich and the other for the poor, might solve the equity problem. It would, however, render the market impotent to do what it does best—guide producers and consumers to decisions that achieve the greatest benefits in consumption and efficiency in production.

When OPEC ran the price up in 1973 and 1974, the world's poor and rich alike paid the new price. For the rich countries the higher price was an economic hardship that impelled them to cut planned oil purchases somewhat, to spend a bit less on other goods, to draw down reserves, and to borrow. The poor countries, already heavily in debt and standing on the brink of disaster, faced reducing both oil and other vital imports so much that their development plans were in jeopardy. Paying for only the minimum required imports plunged these countries deeper and deeper into debt. OPEC's deed annoyed the industrial countries and forced them to make uncomfortable adjustments. To the less-developed countries, it threatened economic and political instability.

The reaction of the poor countries was unexpected. Instead of making common cause with industrial countries to protest the OPEC move, they cheered OPEC and rallied to its side. The universal condemnation of OPEC's act that the industrial countries wanted became instead another cleavage between the rich and the poor. The rise of OPEC forged an alliance between the oil rich and the poor. And in the years since 1973 the alliance has grown. OPEC has nurtured the bond, aided the poor countries, used its oil as a bargaining tool on behalf of all poor countries, and has emerged as their leader.

Despite their recent wealth, members of OPEC are poor countries and have underdeveloped economies. Not one has a diversified and modern economy and all members depend on imports. Not one has a non-oil income per person more than a tiny fraction that of the United States. Not one of them produces anything of significance except oil. They think of themselves as less-developed countries and share the attitudes, myths and realities, and convictions of the poor countries. The other poor countries regard the nations of OPEC as part of the Third World.

THE POOR LOOK AT THE
WORLD ECONOMY

A deep gulf divides how the rich and the poor countries, including OPEC members, view how the world economy operates. The advanced industrial countries see it as harmonious and benevolent, conferring benefits on all participants, with the private market as the central reconciling institution. Although the system has minor imperfections that need repair, no better method for getting maximum production at the lowest prices exists. Tampering with it would only make matters worse. This view derives from the advantages that markets have conferred on developed countries over the centuries. It derives also from the theories of international trade that developed countries have constructed to explain trade and later to support and justify their beliefs in the market's unique advantages.

The poor countries believe the world economy operates unfairly. It is not just a few imperfections but rather basic structural flaws that confer most of the benefits of trade on developed countries with only occasional and uncertain crumbs for the poor countries. They acknowledge that advanced countries have lent them money. They hear the developed countries talk endlessly of their foreign aid. The poor countries, however, regard foreign aid as a palliative that does not address the flaws that keep them poor and in bondage. Only an overhaul of the way that world trade functions will satisfy the less-developed countries by distributing its economic benefits more equitably.

The less-developed countries divide the noncommunist world into two great camps. One camp, consisting of twenty to twenty-five countries, are the developed and industrialized nations. Western Europe, the United States, Canada, Japan, and a few others comprise this group. They even have a club, the Organization for Economic Cooperation and Development. Most are above the equator and less-developed countries call them collectively the North. Their population is modest and grows slowly. Consumption levels are high. Nearly all are highly industrialized. Trade among themselves is extensive. They possess the world's economic wealth and power.

Three-fourths of the world's population, living in more than 120 countries, are the second camp. Consumption levels in these countries, scattered around the world but concentrated in the southern hemisphere, are a small fraction of those in the North. Poverty, malnutrition, and sickness are the rule. Trade among themselves is modest and their total trade is only one-fourth of total world trade. Yet they depend heavily on their imports and exports. Most produce one, two, or three primary products that enter world trade, but the condition of their domestic economy turns on trade. Most of their economies are weak and unstable. These countries, except for OPEC members, have no economic power.

In addition to two groups of countries, the less-developed countries point to two different kinds of markets in the world economy. One set of markets is for trade in raw materials that the less-developed countries export. The other set of markets is for trade in manufactured goods that less-developed countries import. Developed countries import raw materials from the less-developed countries and export manufactured goods to less-developed countries in these two different markets.

The products of less-developed countries can find markets only in the advanced countries where the less-developed countries must buy manufactured goods. Advanced countries, of course, produce and trade in raw materials. Most of the trade in manufactured goods is also among the advanced countries. The less-developed countries thus stand at the periphery of both kinds of markets, with little influence in either. These two sets of markets produce quite different economic results for the two different groups of countries according to the accepted economic thinking of the leaders in the less-developed countries.

Large firms, often international in scope, dominate the markets in manufactured goods. These firms often control much of their own economic environment. They set the price, influence their own demand, must of which is at home or in other high-income countries, and plan their operations, including technological change, far into the future. Their pricing policy emphasizes stability and their fewness guarantees ample profits. Prices seldom go down, although expanding demand and persistent inflation has forced costs and prices up and up.

Profit-taking, expansion of the enterprise, and labor-union demands in the production of manufactured goods in industrial countries prevent declining prices. Even when large-scale production, falling costs, technological improvements, and limited competition would seem to require price reductions, prices remain the same. Labor, management, and the government share the benefits of stable or rising prices.

The raw materials used by the large firms come not only from the less-developed countries but also from developed countries. The enterprises often own or control their own raw-material sources, buying preferentially and always endeavoring to keep raw-material prices low, as the oil companies kept the crude-oil price low for decades. At every opportunity, in order to control more of the production process, they employ technology to reduce the raw-material component of their final products and to replace raw materials with man-made materials.

As the incomes of the buyers of manufactured goods increase, so also does their demand for those goods at an even faster pace. The industrial countries feed one another continued growth through increasing imports and exports of manufactured goods. If the less-developed countries are to develop, they also need increasing amounts of manufactured goods, which helps the growth of already developed countries. Their rapidly growing demand for such imports outstrips the demand for their raw-material exports that grow at the slower rate dictated by buyers in the industrial countries.

Raw-material markets for the poor countries often are not helpful to the poor countries. They are, unlike manufactured-goods markets, highly competitive. A large number of countries, developed and less-developed, produce raw materials. Within each a very large number of individual producers compete to sell their products. No producer and in most cases no nation has much influence in the market. Buyers, most often the large firms in the North, encourage competition to keep supplies ample and prices low. These buyers often engage in raw-materials production themselves in order to guarantee supplies and keep prices down.

The amounts demanded and supplied of raw materials is often

insensitive to price. In the short run, before buyers can find a substitute or redesign the product to reduce the raw-material content, a change in market conditions causes wide price fluctuations. A bumper crop of coffee in Brazil thus not only depresses the coffee price but also reduces the income of all coffee producers. A bad crop year increases price as well as the income of producers.

The demand for raw materials grows slowly. Some raw-materials markets—for example, most foodstuffs—grow only as fast as population grows. Since buyers in the industrial countries are constantly seeking and finding substitutes and ways to use less raw materials, the demand for raw materials grows more slowly than does the income of the buyers. This means that the exports of less-developed countries inches forward at a snail's pace.

When less-developed countries try to break free from the grip of raw-materials production and exports, they confront fierce competition in manufactured goods. The large firms can undersell the products of less-developed countries in the latter's own markets. When these countries use tariffs to protect their home industries, firms from industrial countries buy out local producers. Home production of manufactured goods by national enterprises is still-born.

When producers of the few manufactured goods in less-developed countries try to sell in the markets of the advanced countries, they confront trade barriers. To maintain their own home production, advanced countries levy tariffs against foreign manufactured goods. Raw materials usually come in free. The tariffs apply to all foreign production, regardless of origin, but they damage most the less-developed countries since they are high-cost producers. Poorly experienced in manufactured-good production and sales abroad, the less-developed countries can't get a foot in the door. In the fifteen years before the mid-1970s, less-developed countries did increase their share of world manufactured-goods exports from 5 to 7 percent. Their share of raw-materials exports, however, is 51 percent of the world total.

The picture painted by the less-developed countries depicts unconscious exploitation. They sometimes accuse the advanced countries of deliberate manipulation. More often, they believe that it is the system that consigns them to raw-materials production, fluctuating and often depressed prices, and demand for their products that lags far behind their demand for manufactured goods. At the same time if they are to progress, they must buy at high and rising prices, increasing amounts of the manufactured goods that they are prevented from producing. Locked in these twin vises, the less-developed countries see only a dim future for themselves so long as the present system prevails.

The foreign aid rendered by advanced countries has not relieved the exploitation. Indeed, it may accentuate it. It leads to increasing debt to pay for the North's manufactured goods, a debt that poor countries have difficulty paying because their low-priced raw materials that grow more slowly in price buy less and less inflation-priced manufactured goods that grow rapidly in price. The aid makes them dependent on foreign firms and technology over which they have no control. The less-developed countries believe the aid has been a handout to salve the conscience of industrial countries. It also serves the interests of developed countries. It is not, they say, an effort of sufficient magnitude to set the poor countries on the pathway out of poverty.

WHAT THE POOR WANT

The poor countries assert their right to a world trading system and financial arrangements that promote their economic development as much as they do that of the developed countries. The present system, they argue, confers nearly all the benefits on the rich. Rhetoric and self-serving handouts by the rich no longer suffice. The present system must be replaced, they say, else in time it will destroy not only the poor, but also the rich.

The battle cry goes up for a "new international economic order." These words, first uttered in a conclave of less-developed countries in Algiers just a month before OPEC took control of the price of crude oil, summarizes the reforms that the poor countries propose that the rich countries implement. Progress has been halting, not only because of many technical difficulties and deficiencies in the proposals of the poor countries, but also because of the hostility of the developed countries. Their desperate plight, as well as the opposition of advanced countries, has unified the less-developed countries. They continue to urge their reforms.

What exactly do the poor countries want? They have an eight-item shopping list:

1. More opportunity to sell their manufactured goods in industrial countries.
2. A greater share in the world's manufactured goods production.
3. Higher and more stable prices for their raw material exports.
4. Debt relief through renegotiation of loans.

5. A code regulating the behavior of large multinational corporations.
6. Greater access to modern technology at lower cost.
7. Accelerated foreign aid on favorable terms, mainly through international organizations.
8. A greater voice in international economic and financial decisionmaking.

The less-developed countries want to industrialize. With industry based first on their own raw materials, they then wish to develop the skills and acquire the equipment and technology to produce increasingly complex manufactured goods. At first, production may be high cost and difficult to sell. With assistance from industrial countries and modern technology, at costs that do not price their production out of the market, they can get a start.

The less-developed countries want to sell manufactured goods to developed countries, where the large markets are. Trade barriers now keep them out. So they press for preferences, that is, lower tariffs for their production than those for the products of developed countries. Modest success has attended their efforts. Today 19 developed countries offer limited preferences to 126 countries. The United States in 1976 was the last major country to offer preferences. Most preferences have limitations. The United States, for example, excludes such industries as textiles and shoes, hard-pressed industries at home. These, of course, are precisely the industries in which less-developed countries must begin. The United States also wants the automatic expiration of the preferences in a few years.

Poor countries still need technical know-how. Firms owning the technology prefer most to sell products. Next, they would prefer to own the plants making the products in less-developed countries. Still further down their list of preferences, they would like to sell the plants. Finally, they will rent the technology. Leasing technology is costly and still leaves the large corporations in control, the poor countries say. They want to be free from the technological dominance of foreign firms. Negotiations are difficult. A major obstacle is the control of technology by private firms while the negotiations are between governments.

Technology is only one bone of contention between the less-developed countries and foreign firms. These countries want to prevent or limit foreign ownership and control of domestic firms. They are uneasy about the great economic power of foreign enterprises that often have advantages over domestic producers and can take advantage of their superior position. The less-developed countries want an international code of behavior for the large multinational corporations that will limit their actions. The developed countries, they argue, must enforce the rules, since the less-developed countries cannot make the large firms behave. Without an effective code that is implemented, the less-developed countries fear that they will remain the prey of the powerful.

Less-developed countries realize that their principal role for many years to come will be that of producers and exporters of raw materials. To permit that production to be more beneficial, less-developed countries want higher and more stable prices. They want a large fund, raised primarily by the advanced countries, to guarantee the prices of a group of raw materials by buying when world demand slumps, putting a floor beneath their economies. They also want raw-material prices to reflect the real scarcity of raw materials that they believe to require prices higher than those set by the market. They have suggested that the prices of raw materials move in correspondence to the prices of manufactured goods. In a word, they want a large stabilization fund, parity pricing, much as farmers get in most industrial countries, and artifically dictated prices for some raw materials.

The less-developed countries feel that it is the unfairness of the system that has forced them into debt. The non-oil-producing countries owed $172 billion in 1976 with an annual debt service of $26 billion, nearly 20 percent of the value of their exports. Many countries simply cannot pay. They borrow more to pay off debts and even in some cases to pay the debt service.

Because the system is responsible for their plight, they seek debt relief. They want loans renegotiated, some debts lightened and the terms lengthened and made more lenient. Some debts, especially those of some of the poorest countries, should be forgiven. Industrial countries are listening attentively because they fear a massive default that could hurt them if something isn't done. A 1978 conference made progress not only in recognizing the problem, but also in demonstrating the willingness of some industrial countries—Sweden, for example—to make concessions.

Foreign aid remains essential. The poor countries want more. They want a fixed proportion such as one percent of the gross national product of the developed countries. In addition, they want foreign aid that is independent of the whimsy of national administrations and legislatures. They feel the World Bank, regional banks, and other international institutions, in which they have some say, are more appropriate and less susceptive to political influence.

The less-developed countries want also to receive the benefits of any unearned bonus of the earth. When the International Monetary Fund created out of thin air new money—the SDRs—in the early 1970s, the less-developed countries felt that they should receive it. The developed countries need not worry; it would all end up in their hands when the poor countries spent it. The less-developed countries also want to be the recipients of any benefits from resources not already owned, such as the ocean's mineral resources. Negotiations on the sea's resources continue.

Most of all, the less-developed countries want a louder voice in international economic affairs. Today, a few countries and their institutions control nearly everything—trade, investment, finance— and decide whether or not the poor countries will develop or stagnate. In the World Bank and the IMF the influence of less-developed countries is negligible. Although they host many large companies and are important suppliers and customers, they do not have any influence on the policies of the enterprises. Industrial countries, they say, ignore the interests of poor countries in serving their own. These poor countries believe that they have the right to help make decisions that determine their fate.

POSTURE OF INDUSTRIAL COUNTRIES

For the developed countries to do what the less-developed countries want would cost the developed countries money and would diminish their influence. The leaders of advanced countries, most of them democracies, feel that they would betray their peoples if they did not extract benefits from the proposals equivalent to the loss of money and power incurred in their implementation. Not only do the industrial countries not perceive any benefits for themselves in the new proposals, but also they do not believe that the proposals will work. They see the new international economic order as a romantic and quixotic notion that would lead to disaster.

Much of the divergence originates in the different view of how the world economy operates. The view of the developed countries is older and is sanctified by long-accepted theories. It emphasizes the harmony of interests among nations. If each country produces that which it can produce most cheaply relative to other things it could produce, and then exports these goods, while importing goods similarly produced in other countries, everybody benefits most. No other distribution of the world's production among nations permits all nations to consume more at lower prices. Few indeed are those in industrial countries who do not accept the theory underlying the law of comparative advantage.

Although comparative advantage does not insure that the standard of living is equal among all nations, it does assure that nations employ their resources most efficiently. Its operation should also tend to equalize payments to labor and other productive factors. It has been easy for economists in the North to overlook the theory's strict assumptions, such as perfect competition, free trade, and countries of comparable size and the same stage of development. It has been just as easy for economists in the South to overlook the absence of incentives and pressure for efficiency in their position. So, just as the North gouging the South is an article of faith to less-developed countries, the universal benefit of comparative advantage is a solid conviction among the developed countries.

Advanced countries look askance at what they consider the naive notions of less-developed countries in much the same way that the latter criticizes the self-serving ideas of the rich. The poor argue that the rich hide behind comparative advantage because it justifies what the rich do. The poor argue that comparative advantage applies only in a world in which no one possesses any more power than anyone else—a flat contradiction of the facts. They also point out that the comparative-advantage world is a changeless world—no technological change, no depleting resources, no new resources, no changes in demand. Comparative advantage, they say, applies to an ideal world that exists only in the economics textbooks.

The belief in the developed countries that the advantages of the market system for everyone lingers on. Buttressing this belief is the fact that the present methods of world trade and finance do work and do serve the interests of the advanced countries. Unlike the less-developed countries, the developed countries do not believe the patient is gravely ill. Trade and finance may be a bit sick, perhaps, and require some attention. The radical surgery recommended by the less-developed countries, however, will not cure the patient. Indeed, it could kill him.

Disputing the diagnosis of the less-developed countries, the economic doctors of the advanced countries have zeroed in on the weaknesses of the specific remedies and the advantages of the actions already taken. Item by item they tick off the new ideas showing that the proposals won't work, that present arrangements are better, and that measures short of radical change have permitted trade to make a substantial improvement in the situation in less-developed countries. They argue that whether or not the poor countries develop depends less on the conditions of the international economy than it does on the initiative, frugality, and hard work of the poor countries.

The advanced countries believe that unless accelerated industrialization in poor countries reduces costs enough for them to compete with enterprises in advanced countries, they cannot industrialize. Cost reductions depend on what each country does at home, not on changes in the world economy. It is a matter of efficiency, skill, and costs. Advanced countries often neglect to mention that their restrictions on imports also raises the costs of domestic manufacturing in less-developed countries.

The prices of raw materials depend on their supply and demand. A fund may help to stabilize the price by buying during flagging demand and selling during active demand. The rich countries have indicated that they will participate in such funds, as they do now for some raw materials. They prefer a system in which countries can receive assistance when the value of their raw-material exports go down seriously and then repay when export values improve. Such a method is in place in the European Economic Community for many less-developed countries associated with it. The method leaves price alone but tries to repair the damage done by price changes.

The rich countries are convinced that no stabilization plan can raise raw-material prices above the level dictated by their markets. A price higher than the market price calls forth additional supplies and diminishes demand. Eventually, no matter how large the fund, it would run out of money to buy the new supplies that no one else wants to buy. The new supplies would bring unbearable pressure to bear on raw-material prices.

Only by curtailing supply, as OPEC does, creating an artificially high market price, could the fund stay in business. But what country will restrict its supply? Everyone wants everyone else to cut supplies. Cooperation among producers, in the OPEC fashion, might restrain production, but such cooperation is extraordinarily difficult to establish and maintain. Such cooperation will work only under highly specialized conditions. The advanced countries point out that they export nearly four times more raw materials, excluding oil, than do the less-developed countries. The plan would benefit the rich much more than the poor.

Economists in advanced countries ridicule the notion of indexation, tying the prices of raw materials to those of manufactured goods. These are two different sets of markets, with different supply and demand conditions. Raw materials are only a small part of the cost of finished products. Raising the price of iron ore because auto workers in Germany, in negotiating a more favorable wage, forced auto prices up, makes no economic sense.

The glut of iron ore would soon burden the market. Its price would have to fall without an increase in ore demand. Workers in Germany, however, will buy little iron ore with their pay increases. Advanced countries concede that many markets are imperfect. But the suppression of raw-materials markets, replacing them with formulas and bureaucracies, would result in chaos.

The debt owed by the poor countries is the savings, through taxes and voluntary actions, of people in advanced countries. These investors, who are responsible for those savings, negotiated loans and their terms in good faith and are entitled to repayment plus interest as agreed. It is unfair to them to cancel the loans or renegotiate the conditions. In addition, renegotiation will certainly not encourage savers to lend any more, a potential loss to the borrowers. Advanced countries recognize that many borrowers are in such desperate straits that renegotiation in some cases would be better than default. Even though accepting limited renegotiation, advanced countries urge the poor countries to tread carefully lest the sources that provide the loans dry up.

The advanced countries do not accept the proposition that the less-developed countries have any special claim on their savings. Savings flow into those economic activities and to those places in the world that are safe, productive, and pay adequate returns. When less-developed countries meet these conditions, they will acquire more capital. Recognizing the special problems of poor countries, advanced countries have mounted many public programs of loans and grants, some national and some international.

The advanced countries deny that the amount of foreign-aid programs should be some arbitrary percentage of their production. This means that countries who cannot produce competitively receive a reward because other countries do produce more goods efficiently. If these programs are to become an obligation of the rich, then the poor must accept the obligation to use the savings effectively. To assure that the less-developed countries are discharging this obligation would require supervision unacceptable to any nation. Formulas that turn money over to the poor simply because they are poor, the industrial countries argue, provides no incentive to improve.

The behavior of the large multinational corporations often justifies the complaints against them. These enterprises tend to do what they can get away with. Their more tempered actions in their home countries do not reflect a different attitude at home than abroad. It reflects the laws and constraints at home that they do not confront abroad. The recent scandals of bribery, corruption, and dirty money in the United States indicate that the multinationals will skirt any law anywhere and break it if they feel they can escape retribution.

With adequate countervailing power, the governments of advanced countries try to hold the large companies in check. They argue that the poor countries should do the same. The multinationals will behave when governments set unambiguous rules and enforce them. An international code will work only when the individual governments dealing with multinational companies accept and adhere to it as their own. No code is needed if each country exercises its own sovereignty and controls the multinationals under its jurisdiction. If the less-developed countries want a code, then it falls to them to develop and enforce it.

The large companies spend billions of dollars each year developing new technology. They do it because they expect to receive economic benefits. The companies would let technology stagnate unless they receive a payoff. The advanced countries want to encourage the transfer of technology, but fear that technological development will dry up if the companies do not benefit and must give technology to the less-developed countries. These countries would also enjoy an unjustified cost advantage unless they pay at least a part of the cost of developing new technology.

New technology reflects the availability of labor and capital in the countries developing it. Since the United States, Western Europe, and Japan are responsible for most new technology, it reflects the large amount of capital and scarcity of labor in those countries. The poor countries have a great deal of unskilled and semi-skilled labor but capital is scarce. To introduce the technology produced for one type of economy into another type distorts the latter and can lead to the waste of capital and to unemployed labor. More than new technology, the less-developed countries need the ability to create their own technology based on their resource endowment. Such an ability comes from education, experience, and the determination to do it, not from copying advanced countries. Technology is a domestic challenge.

The rich countries have no wish to stifle the voice of less-developed countries in international councils. They feel that the organization of world bodies—the United Nations, World Bank, International Monetary Fund, and many others—guarantee that less-developed countries have ample opportunity to state their views. Most of these organizations—the United Nations General Assembly is an exception—are based on participation and contribution, not mere existence. The World Bank and many others measure the voting of members by their contribution to total production, trade, or financial assets.

To give a small country whose contribution to production is 0.1 percent a voting weight to 10 percent means to deny another country whose contribution is 10 percent its legitimate place. In deciding on world economic and financial matters affecting all countries the only fair arrangement is to give each country a decisionmaking influence equivalent to its share of goods produced or some other measure of its contribution. Germany has more to lose than Haiti and international economic organizations must recognize its greater stake. Even the United Nations recognizes the importance of the great powers in its Security Council.

The main point that the developed countries make is that development of the less-developed countries is not an international problem but rather comes or does not come from within. International trade and finance condition development, helping or hurting it a bit. Whether or not the poor countries progress depends on their own efforts, skills, industry, and the efficient use of their resources. The rich countries can cheer from the sidelines and send in a few plays but they cannot play the game. They recognize that the international economy on occasion penalizes the poor countries, as it also does the rich countries. As these cases occur, the advanced countries will adopt policies to improve the system. But to dismantle and remake the system serves no one.

IMPACT OF OPEC

The grumblings of the poor and the protestations of the rich by the 1950s and 1960s had become standard world rhetoric. Leaders and officials of less-developed countries made their points at great length at every opportunity. Representatives of developed countries answered at equal length. The two groups talked at one another, with seldom a meeting of the minds. The rich made loans and grants and were unhappy when the poor did not show the appropriate gratitude. Even when the rich told the poor exactly how to spend the money and provided the money, these countries didn't improve much. The poor were divided among themselves and each spoke only of its special problems. But the poor countries were beginning to perceive the advantages of cooperation.

As early as 1961 a large number of less-developed countries met under the banner of the Group of Non-aligned Countries. Its purpose was political, to avoid being dragged into conflict as pawns in the cold war. Initially, the Group ignored economic matters. As the 1960s drew on, cooperation for economic purposes became more important. Largely through the efforts of Dr. Raul Prebisch, an Argentine and an international civil servant, the United Nations created the United Nations Conference on Trade and Development (UNCTAD). It has become the U.N. "club of the poor."

UNCTAD, whose membership consists exclusively of the less-developed countries, dealt with their economic problems. Their periodic meetings provided the less-developed countries with a platform from which to make their charges against the advanced countries and to extol the virtues of cooperation. By 1970 when the United Nations Decade of Development, which witnessed only a small improvement in the plight of less-developed countries, closed, the theme of economic development had the highest priority. The strategy and outlook of the poor countries took form.

The poor countries formed the Group of 77, an informal alliance now embracing 117 countries. It remains their most effective organization. It wields no power, but along with OPEC, UNCTAD, and the UN General Assembly, it voices the economic concerns as the pressure group of the less-developed countries. The exchange of ideas and information in the Group of 77 produced their analysis of the world economy, their view of what must be done to change it, and the strategy of bargaining to achieve their goals. Still, in the early 1970s the poor countries did not have enough influence to persuade the rich countries to the bargaining table, much less to force concessions.

All that changed suddenly in 1973 when OPEC captured the crude-oil price. Three things happened at once. The economic position of many less-developed countries worsened since they had to pay four times as much for oil. The developed countries, consuming vastly more oil, had to pay much more and to a group of less-developed countries. The third was the realization that even less-developed countries can, through cooperation, significantly improve their development prospects and make the developed countries bow to their will.

Most less-developed countries were already going into debt every year as import needs for development exceeded their raw-material exports. At the end of 1972 they owed about $60 billion, accumulated over a long period of time. In the years since, the debt has shot up under the impact of higher oil prices and the induced recession in the developed countries. The debt at the end of 1977 was over $200 billion and the annual payments are nearly half of their total debt five years ago. Debt threatens the stability and undermines future credit. The less-developed countries feel that their worsening position makes reform urgent.

On the basis of the OPEC price rise, the less-developed countries should have been irate. They were not pleased, of course, to pay the higher price. It hurt them even more than it did the United States, Europe, and Japan. Logic should have compelled them to make common cause with the advanced countries against OPEC. But they didn't. Indeed, they did just the reverse. They allied themselves with OPEC and hailed its victory as their own.

The less-developed countries were aware of the desperate condition into which the oil-price rise plunged them. Some countries suggested to OPEC a two-tier pricing system, one price for advanced countries and a lower price for less-developed countries. OPEC responded that oil is oil and discrimination by customer or geographic area would not work in the oil market. Transhipment and diversion would undermine it. Some countries suggested that OPEC make loans to offset the price increase. In 1974 OPEC members initiated their foreign aid programs and assured the less-developed countries that they would help in every way possible.

The attitude of the poor countries reflected their growing solidarity. The voice of the Group of 77 became more strident. Increasingly, the less-developed countries spoke of their economic rights. The United Nations General Assembly—where one nation-one vote prevails—became a battleground. The less-developed countries pushed through in 1975 the Declaration of Economic Rights which reflected their thoughts on world economic reform.

The less-developed countries envied and wished to emulate OPEC. Any group of less-developed countries that successfully twists the tail of the rich lions of the world receives their admiration, even though the swishing tail staggers them as well. They were quick to see in 1974 the historic significance of OPEC's move. Never before had a group of less-developed countries used their combined economic power to force concessions from the advanced countries. The lesson of OPEC—that cooperation pays—was learned well.

Among the responses of the advanced countries to the oil-price rise was a call to OPEC to meet and discuss world oil. Naturally, out of a meeting between OPEC and advanced oil-consuming countries, the United States and others hoped to persuade OPEC to follow a moderate course. OPEC, having just won the battle, saw no compelling need to meet. OPEC did see some advantage in a meeting of less-developed and developed countries to discuss oil as well as all other economic matters. Oil might get the advanced countries to the bargaining table.

When President Giscard D'Estang of France invited oil importers and exporters to meet in Paris, the conference quickly resolved into a meeting between developed and less-developed countries. At the preliminary meeting the United States and others insisted that they talk only about oil. The less-developed countries, encouraged and abetted by the OPEC members in their midst, said the meetings must concern all economic development problems. After much talk and no agreement on what to talk about, the meeting collapsed.

Within months the advanced countries reconsidered. They reached a compromise that was a victory for the less-developed countries. The work of the meetings would be in four committees. One committee would consider oil. The other three would examine other development problems. Each committee and the plenary meeting would have two chairmen, one from the less-developed countries, one from the advanced countries. Twenty-seven nations sent representatives to the Conference on International Economic Development, unofficially dubbed the Paris North-South Dialogue. Dr. Manuel Perez Guerrero of OPEC's Venezuela chaired the meetings for the less-developed countries. The meetings began with high hopes and expectations. Something at last would happen.

Despite the support of the UN General Assembly Declaration of Economic Rights and the support of all but the advanced countries, the less-developed countries in Paris received little satisfaction. The end of one administration in the United States and the beginning of another caused some delay. The kinder words and warmer sympathy of the Carter administration, however, did not presage any greater concessions. The meetings dragged on.

The advanced countries stalled on two grounds. One was economic self-interest. They knew that what the less-developed countries were proposing would cost money and would cause worrisome adjustments at home. They also had grave misgivings about how the proposals would work. They feared that many of the plans would backfire and worsen the conditions of trade, aid, and investment, and would even damage the interests of the less-developed countries as well as their own.

Despite the urgency felt by the less-developed countries, the discussions went on and on. Participants talked at and past one another. Domestic problems and other international issues, including touchy armaments, the East-West conflict, the Middle Eastern and African troubles, and energy, environment, and the economy at home commanded much of the attention of the developed countries. Finally, by mutual agreement, the meetings adjourned in mid-1977. Bitter, the less-developed countries took their case back to the United Nations.

The fight continues. In the committee of the whole—all 150 countries—of the United Nations, the Group of 77 vows to press for reform. The developed countries continue to resist. In May 1978, the committee adjourned with no progress. Part of the problem in this round was a difference in interpreting the role of the committee. The less-developed countries insist that the committee has political power. The advanced countries regard it as a debating forum.

The fifth meeting of UNCTAD in Manila in May 1979, bore small fruit. Industrial and socialist countries attended, and the less-developed countries hoped to secure agreement on the financing of commodity stabilization and on other matters for which they had been pressing. Debate was intense with the usual charges and countercharges hurled about. The meeting reached some consensus agreements—non-binding—on issues that do not touch the core of the problem—the transfer of resources from developed to less-developed countries.

The conflict between the two groups of countries will not go away. The developed countries have made it clear that they will not accede to the demands as they stand, although they have made some concessions. Over a long period of time, the industrial countries may give ground under hard pressure and through bargaining that modified the position of the less-developed countries as well. So far, the less-developed countries show no inclination to back down, hoping that through continued pressure, the help of OPEC, and the justice of their position, the advanced countries will eventually yield. It seems improbable that real progress will be made until the bargaining position of the less-developed countries, aided by OPEC in command of a chronically tight oil market, improves.

The possibility of an all-out confrontation at some time exists. If the less-developed countries lose all hope in bargaining to get their demands, they might try an economic embargo and boycott, nationalize foreign investments, and disrupt world trade. The outcome of a confrontation is uncertain. The greater power, of course, lies with the developed countries, but the disruption, especially if OPEC uses the oil weapon, could devastate both sides with armed conflict a possibility.

OPEC as an organization has not directly injected itself into the North-South conflict. The attitudes of the secretariat, however, are plainly visible and OPEC's publications support the view of the less-developed countries. Members of OPEC as countries participate in meetings of less-developed countries and use their position to exercise leadership. A few take the more astringent position that OPEC's business is oil, not meddling in world politics. Even so, most have made it clear that they support the goals and proposals of the less-developed countries.

President Carlos Andres Perez of Venezuela told the United Nations in late 1976, "We, the countries belonging to OPEC, have made the historic opening toward a new power of negotiations, which is in the hands of Third World countries for the first time. The price of oil will stop increasing when a balanced system is created in world trade relations." (His speech appears in *Documentación*, Venezuelan U.N. Mission, New York, vol. II, no. 9, November 30, 1976, p. 42.) Although it is unlikely that OPEC would use its oil on behalf of the poor countries except in an all-out North-South confrontation, members will act as spokesmen for poor countries and will use their oil as a bargaining instrument.

OPEC ASSISTANCE

OPEC helps the less-developed countries in three ways. It champions the cause of these countries through public and private support, through bargaining on their side, and through pressure on the advanced countries. It also serves as a model for similar export organizations already started or being considered by less-developed countries. Finally, it provides financial assistance through foreign-aid programs.

Without OPEC support the discussions between the developed and less-developed countries would never had occurred. The advanced countries would simply not have come to the bargaining table. The poor countries would still be fulminating in vain. That the advanced countries at least give lip service to changes in the world economy that will improve the lot of the poor countries is a tribute not only to the steady pressure of less-developed countries, but also to the cudgel that OPEC wields. That only limited progress has crowned the efforts is attributable not to the failure of OPEC support but rather to entrenched interests and the extreme complexity of the problems.

OPEC has also presented the less-developed countries with a model of economic organization. If a limited number of countries control a significant portion of the supplies of an important raw material in

demand in advanced countries, then cooperation may yield benefits similar to those of OPEC. Raw-material exporters have met and
endeavored to work out such cooperative agreements for many raw
materials. In some cases, OPEC members have participated in the
discussions. So far, however, national interests and market conditions different from those of oil have prevented a proliferation
of export organizations.

Bauxite producers joined in the International Bauxite Association
in 1973 to raise the price and limit supplies. Eleven countries, all
major producers, of which Jamaica is the most aggressive, have had
some success. Iron ore exporters have also formed an association.
In formation is an Intergovernmental Council of Copper Exporting
Countries, instigated by four major producers. The International
Tin Council includes a price-stabilization scheme. The International
Coffee Organization has both importers and exporters in a stabilization plan, setting export quotas for members. An Organization of
Banana Exporting Countries, including the four Central American
producers, have set a common export tax. Negotiations go forward
for nickel, rubber, tungsten, molybdenum, and other raw materials.

An export organization that hopes to achieve benefits similar to
those of OPEC must be able and willing to control a large portion of
total world exports of the raw material. Otherwise, rapid expansion
of production in other countries will undermine the agreement.
The exporters must be willing to curtail their sales to maintain
the market price. The raw material must have no ready substitute
that can supply the buyers' needs. Reproducible raw materials,
such as foodstuffs, textiles, and forestry products, do not fill the
bill. An exception might be some tropical products such as bananas
and coffee. Even those products, however, face substitutes.

Unique conditions must prevail on the demand side. Price must
be highly sensitive to changes in supply. When supply declines by a
given percentage, the price must go up by a greater percentage.
Otherwise, a cut in the amount of supplies will result in a loss of income, rather than an increase. No benefit results if the percentage
cut in price increases the amount demanded by the same percentage.
More is sold but the earnings are the same. Benefits to suppliers
also depend upon how fast demand grows; bouyant demand helps.
The slower growing demand for foodstuffs, geared to population
increases, offers little prospects for benefits from cooperation.

The kinds of products that can yield great benefits are those
that are necessities, have no substitutes, and whose supply is controlled by a few countries. Oil is almost unique. Some metals such
as aluminum and iron ore may qualify. The list, however, is short.

Much depends upon the willingness of exporting countries to control supply effectively, setting it at a level that will maintain the desired price, as well as the response of importing countries. If importing countries can reduce demand efficiently, they can frustrate the exporters' best laid plans.

The International Bauxite Association and the Iron Ore Exporters' Association have met with modest success. Each contains the major producers and can influence the price heavily by regulating their supplies. The price of both went up in the 1970s and have remained relatively high with increased earnings for the exporting countries. In neither case have the benefits been as spectacular as those of OPEC. Market conditions were just not as favorable, and they are but pale copies of the market conditions that OPEC enjoys.

The success of OPEC has touched off a renewed interest in all types of commodity agreements and producer cooperation. Many plans also include importers as well as exporters, plans that stabilize price, but that do not contemplate raising it artificially. Tin, wheat, coffee, and other commodity agreements function by buying in times of slack demand, accumulating a stockpile, and then selling from the stockpile when demand is brisk, keeping the price within a limited range. The proposal of the less-developed countries at Paris was a Common Fund for many products, which would economize on the money needed. So far, however, indexation or any plan to raise prices above the market level has gained no adherents outside the less-developed countries.

The large monetary assets of OPEC members provide the basis of the third type of assistance—foreign aid. Never before had one group of less-developed countries made an effort to assist another group of less-developed countries with financial resources. OPEC did. It began even before the price revolution of 1973, but only in the last few years has it become significant. The OPEC foreign-aid experiment is the subject of the next chapter.

Chapter 9

OPEC Foreign Aid

Since the oil-price revolution, OPEC members have given or lent at least eighty other less-developed countries between $45 and $50 billion. In 1977 these foreign-aid programs were two and one-half times greater than that of the United States and nearly two-thirds the amount of aid of the seventeen members of the Development Assistance Group (DAG) of the Organization of Economic Cooperation and Development (OECD). If the DAG countries had given assistance in the same ratio of aid to gross national product as did OPEC members, they would have given $175 billion instead of the $18 billion they actually provided.

In just a few years OPEC aid programs have become a significant element in total world development assistance. OPEC members have displayed a greater willingness, relative to resources, than industrial countries in providing the less-developed countries with financial support. Despite the fact that members use surplus oil revenues to sustain their aid, the same funds could earn more with less risk if put to other uses.

Before 1973 OPEC foreign aid was modest. It consisted primarily of assistance by Arab OPEC members—mainly Libya, Saudi Arabia, and Kuwait—to other Arab countries, especially Egypt, Syria, and Jordan. When the price of crude oil rose, nearly all OPEC members immediately began to dispense aid. They began first in neighboring and allied countries, but gradually broadening into a worldwide program. When the Sovereigns and Heads of State of OPEC members met in Algiers in March 1975, they pledged themselves to accelerated aid to less-developed countries. In mid-1976 the Opec Special Fund began operations to supplement the bilateral and multilateral aid of individual members.

Although OPEC aid programs are new and modest, they signify the willingness of members to participate fully in financing in development assistance. The Arab Gulf countries and Iran have been the principal contributors but even the smaller and hard-pressed members have made a showing. Although OPEC cannot and will not take over the financing of development in less-developed countries, its programs indicate that within the limits of financial feasibility, OPEC members are backing up their moral and political commitments to other less-developed countries with money.

DEVELOPMENT OF OPEC
FOREIGN AID

To the tiny sheikhdom of Kuwait belongs the distinction of initiating the first OPEC aid program. In 1961 that country established the Kuwait Fund for Arab Economic Development. In its first twelve years of operations, it lent, on easy terms, nearly $500 million to twelve Arab countries. Kuwait insisted on high loan and performance standards and has built a small but effective system for granting foreign aid. Nearly one-half of the more than forty loans up to 1973 were to Egypt, Sudan, and Tunisia. The Kuwait commitment constituted 5 to 6 percent of its gross national product. Much of the later outlook of other OPEC members has been influenced by the Kuwait experience.

On the heels of the Arab-Israeli war of June 1967, three OPEC members—Saudi Arabia, Kuwait, and Libya—established a support program for the "front-line" Arab states of Egypt, Syria, and Jordan. In Khartoum, in August 1970, the oil states agreed to provide an unspecified amount of budgetary support for the belligerents and the countries bordering Israel. The amount, which was in form of grants, not loans, averaged $400 million a year between 1970 and 1973. It was supplemented by other financial assistance to buy arms. This aid, avowedly political and military, is not usually counted in OPEC aid figures. Nor do the donors regarded it as a part of their aid programs. This silent support, however, has been of critical economic importance to the recipients.

Arab OPEC members also contributed to the founding in 1968 of the Arab Fund for Social and Economic Development as an Arab League institution. Other Arab countries made modest contributions. Its purpose, like that of the Kuwait Fund, was to make loans on concessional terms to the least prosperous of the Arab countries. In 1971 the most important component of the United Arab Emirates established the Abu Dhabi Fund for Arab Economic Development, again patterned after the Kuwait experience. Venezuela contributed to the Andean Development Corporation. All of the aid of OPEC members before the price revolution was closely tied to the foreign policy goals of the donors.

All of the aid of OPEC members from 1970 through 1972 amounted to only $1.8 billion. Nearly three-fourths of it consisted of bilateral loans on easy terms in the Arab Middle East, although there were some regular loans, as well as a small amount of multilateral aid. A small amount also went to the Moslem countries in Africa.

In 1973 OPEC members accelerated their foreign aid. In that year aid was of about the same magnitude as the total of the three previous years—$1.7 billion. Nearly one-third of it came from Kuwait. Over 90 percent of the aid came from Kuwait, Libya, Saudi Arabia, and the United Arab Emirates. Again, most of the funds were in the form of concessional loans by the Arab OPEC members to other Arab countries. All OPEC members, however, had some foreign aid.

Through the end of 1973, OPEC members had given no serious thought to a worldwide foreign-aid program. The Kuwait and Abu Dhabi Funds, as well as the regional Arab Fund, reflected the brotherhood of Arab countries. Aid was for friends and neighbors. Although the transactions, except for the front-line silent aid, proposed to support the recipient's economic development by assisting in specific projects, they were not a part of a deliberately designed program by OPEC members to assist non-OPEC less-developed countries. Indeed, the fact that the aid donors were members of OPEC was incidental.

OPEC OUTLOOK ON AID

Not long after the spectacular events of the fall of 1973, OPEC members began to think consciously about foreign aid. In a short period of time, they had to go through all the issues and controversies that had concerned the developed countries for decades. They did have an advantage. Not long before, some of them had been recipients of aid. Multilateral versus bilateral aid, project aid versus general support, as well as the magnitude, institutions, means, motives, and many other matters had to be sorted out. Some are still pending, but the basic outlook and policies of OPEC on matters concerning foreign aid have now emerged.

The first question, whether or not to give aid, never seems to have arisen. National leaders of OPEC members, sensing the critical need of less-developed countries for aid, immediately voiced support for financial assistance. OPEC also sensed that in order to earn and retain the support of less-developed countries and to assert its leadership among them, it must do something quickly. In that first year, 1974, the aid of OPEC members more than quintupled, to $5.9 billion. All members participated. Only one contributed less than $100 million. Saudi Arabia contributed $1.6 billion. The programs that year were hastily thrown together as the members continued to debate how to handle their foreign aid.

Early in the debate, the members of OPEC repudiated any obligation on their part to make good for the increased price of oil through grants. Some less-developed countries had approached OPEC with the proposition that OPEC should in effect return the money that the less-developed countries had to pay for the higher priced oil. Two methods surfaced. One was the idea of an oil-aid grant, similar to the food-aid grant sometimes used by the United States and food-exporting countries. The other was a two-tier pricing system. The industrial countries would pay one price; the less developed countries would pay a lower price.

OPEC rejected both proposals and even the idea that aid should relate to the price and volume of oil exported by OPEC to the less-developed countries. Such a broad and indiscriminate program, OPEC argued, did not take account of the needs of the recipient or any of the standards for granting foreign aid. Members also argued that two prices in the oil market would not work because of the possibility of transhipment. Finally, they asserted that the notion of tying aid to prices would be unmanageable and would remove the meaning of price changes as an indicator of the scarcity of the product. In effect, OPEC opted for a discretionary foreign-aid program.

In his article in *Foreign Affairs* in January 1976, Maurice Williams, an official of the OECD Development Advisory Committee, the seventeen-nation group that keeps tabs on the aid of industrial countries, argues the case for tying aid to the effects of the oil-price increase on less-developed countries. He cites the $11 or $12 billion added cost in 1974 to the less-developed countries on account of the price increase and the $4.6 billion OPEC aid program that year. The next year, the current account deficit was $10 billion more, and the OPEC aid commitments were only about $6 billion.

Mr. Williams' early data were faulty and although he qualifies his remarks, he implied that the basic problem of the less-developed countries in 1973-1975 came from oil. But in 1973 the less-developed countries already had a large deficit—$12 billion—in their current payments. The oil-price hike added to it, but so did the price increases of other raw materials, as well as the recession in the industrial countries, neither of which were uniquely attributable to the oil-price change. OPEC in fact gave $6 billion in 1974, as well as $1.6 billion more in multilateral aid, mainly through the International Monetary Fund oil facility. In 1975 OPEC assistance was $8.2 billion, as well as $3.4 billion more in multilateral aid, almost double Mr. Williams' estimate.

More important, OPEC rejects completely this line of reasoning. The plight of the less-developed countries, including their own before they struck it rich with oil, was the result of a complex set of interrelated factors, as sketched in the previous chapter. Oil played a minor role in their problems in 1974, as well as before and after that date. OPEC would have no part in aid programs that linked aid and oil together, any more than industrial countries will gear the magnitude of their aid programs to their export prices.

OPEC's basic position is that the less-developed countries need foreign aid for many reasons, most of them related to the structure of world trade and finance that treats them shabbily. OPEC members will, at their discretion, render such aid as they can in a way that they regard as most helpful. In fact, as it has turned out, the amount of OPEC aid since 1973 has been only slightly less than the losses of less-developed countries directly attributable to the oil-price increases.

Early proposals for a grandiose central aid fund also met with firm rejection. In 1974 the Shah of Iran proposed a large fund with contributions both by OPEC members and industrial countries. Neither group of countries displayed any enthusiasm for that idea. The shah, always full of ideas, also proposed an assessment on each barrel of oil sold to establish a central fund. Other members of OPEC also suggested a giant aid fund. President Carlos Andres Perez of Venezuela proposed a grant fund. All were rejected. In effect, OPEC members decided that most of their aid would remain in the hands of members to dispense as each wishes.

OPEC MOTIVES

The question of motives in their foreign-aid programs has bothered OPEC members. Why should they—all less-developed countries—give foreign aid? Unlike the aid programs of industrial countries that contribute significantly to their own economies, OPEC members could derive almost no economic benefits. The members were not participants in the East-West rivalry and had no political axe to grind in that arena. OPEC members were too mixed a bag of countries to have a single political or military foreign-policy goal common to all. Most of the funds could soon be used effectively to finance domestic development programs. Any excess over their own needs could earn more in world capital markets than in assistance to less-developed countries.

Each of the members did have its own political motives and justification for foreign aid. In addition, as a group the members wished to establish solidarity with other less-developed countries to strengthen the third force in the world. As a more powerful force, all the less-developed countries together could effect reforms in the way that international markets operated. The absence of these reforms was deterring not only their progress, but also the progress of OPEC members. Members did not believe that the opposition of the less-developed countries could undo their achievement in capturing control of the oil industry. They did believe, however, that their own economic and political position would gain strength by support of other poor countries and that together all less-developed countries could bargain more effectively with the industrial countries.

Midst the rhetoric of the Solemn Declaration of the Sovereigns and Heads of State of OPEC members in 1975, the clear commitment to provide financial assistance stands out. The context was the effort to cement the union of interests of all-less-developed countries, the recognition of some unfortunate side effects of the oil-price increase—although blaming underdevelopment and lack of progress on the industrial countries—and the willingness to use oil as a bargaining tool on behalf of the less-developed countries.

In effect, OPEC members promised that the events in the crude-oil industry, although they added to the cost of oil, would be to all less-developed countries a long-run net economic benefit. OPEC would enable them all to deal with industrial countries on improved terms. OPEC would assist economically. In return, OPEC asked for the support of other less-developed countries in forming a grand coalition of the nonindustrial world.

The ethical and humanitarian motives of OPEC aid are clear. The largest aid-givers are the Arab OPEC members and the largest recipients are the non-OPEC Arab countries. Even as the program has broadened to include first Africa, then Asia and Latin America, the poorest of less-developed countries have had the highest priority in OPEC aid. Although members share the fear of developed-country donors that these countries may waste the money, the humanitarian pull is too strong to deny them aid.

The economic benefits for themselves, OPEC members argue, are nil. Indeed, members say aid programs cost them. Alternative uses of the funds either in domestic development programs or in investments in world capital markets would yield much higher returns at a much smaller risk. In addition, funds now are more valuable than funds later, else there would be no point in paying a premium to forego present consumption. In addition, the high probability of inflation further deteriorates the value of future funds. Concessional loans, a large part of OPEC programs, are thus doubly costly in that returns are smaller than an alternative return on present as opposed to future financial resources.

OPEC members also point out that the funds that they are using in their aid programs come from permanent assets, not current production. Loans are made on the basis of members' oil incomes that in turn are based on the sale of fixed assets. The aid of industrial countries is based on the use of current production. OPEC aid is the conversion of a capital asset—and a depletable one at that—into monetary form.

Nor does the money come back to OPEC until repaid. In the aid programs of developed countries, many loans and grants are tied to the purchase of products from the donor country. Even funds not specifically tied are often spent in the granting country. These purchases support industry and agriculture in the industrial country giving the aid. The strongest domestic support for the United States aid program, for example, comes from those industries and states whose exports increase as a result of the aid. Some quips have said that the U.S. program is really an aid program for American agriculture and industry. Without these economic benefits, some of the aid of industrial countries would find little favor.

OPEC members are not anxious to increase their oil sales to less-developed countries. Indeed, a part of the purpose of the price increases was to promote the conservation of oil, to make it last longer, and to enable the oil to contribute to the long-run development of member's economies. OPEC members, unable to provide anything the less-developed countries need except oil do not benefit from their aid. They would sell about the same amount of oil with or without the aid.

OPEC members are not slow to point out that they are responsible for some of the aid provided by developed countries as well. OPEC investments in banks and financial institutions permit them to make loans to less-developed countries. The absence of OPEC funds in world capital markets would narrow the financial base from which the less-developed countries can borrow. In addition, since recipients spend their money to make purchases in developed countries, the aid-giving potential of industrial countries increases with OPEC aid. Furthermore, to the industrial countries, OPEC aid is an embarrassment and an incentive to increase their foreign aid programs.

OPEC members seem not to have decided how much to give. Each year each member makes its decision without any fixed commitment as to how much. Clearly, the amount is related to the financial condition of the member, particularly the presence or absence of surplus funds, as well as the member's perception of need. Those with large surpluses give most. But even those with no surplus give some. Some OPEC members are even borrowing from the IMF or other sources in the capital market on stiffer terms than they grant in their aid programs.

ORGANIZATION OF OPEC AID

OPEC has supported existing national and international institutions rather than create a new institution and bureaucracy. All OPEC members except Nigeria, Gabon, Indonesia, and Ecuador have their own national aid-granting institution. The Kuwait and Abu Dhabi Funds have been broadened to supply aid for all less-developed countries. Other members have created funds for dispensing aid or use their central banks. In some cases, these institutions put the aid funds directly in the hands of the recipients.

A substantial amount of OPEC aid funnels through existing international organizations. The International Monetary Fund, the World Bank, the Asian Bank, the Inter-American Development Bank, the International Development Association, the Kuwait Fund for Economic Development, the Arab Fund for Social and Economic Development, the Abu Dhabi Fund for Economic

Development, the Special Arab Aid Fund for Africa, the Saudi Arabian Fund for Development, the Iraq Fund for Economic Development, the Arab Monetary Fund, as well as the Opec Special Fund, the Islamic Development Bank, and the Arab Bank for Economic Development in Africa manage most of the money without any direct connection between donor and recipient except in the case of national funds.

OPEC members have a preference for project aid rather than general development assistance. Members rely on existing institutions because they can undertake the lengthy and costly evaluation procedure, relieving members of the need for establishing a new bureaucracy. By working with others, OPEC members can provide funds in blocs for specific projects but without the problem of evaluating projects and detailed management of the funds.

Through the various national, regional, and international organizations OPEC is supporting hundreds of development projects. For example, the Arab Bank for Economic Development has made forty-three loans from late 1975 to mid-1978 to twenty-six African countries for agriculture, transportation, power, and industry. The Kuwait Fund, the channel for many OPEC operations, is funding 101 projects in forty-four countries. All are long-term loans at low interest rates. The Saudi Fund is building a university in Syria, a dam and water and sewage facilities in Tunisia, a sugar project in Somalia, a highway in Taiwan, a fertilizer project in Indonesia, electric power facilities in India, all at a cost of $1.5 billion. It has sixty-three other projects costing nearly $6 billion.

Still, general economic assistance is an important element in OPEC aid. Recipients can use this kind of support to pay any international debt they choose or can fund new projects requiring foreign payments. The Persian Gulf Arab countries mounted a $2 billion program for Egypt in 1976. Egypt's economic difficulties persuaded them, however, to convert the program to general balance of payments assistance. With either project or general aid, OPEC has indicated that it does not want to become involved in detailed evaluation or supervision of its aid.

GLOBAL RECORD

OPEC members have provided from 1974 through 1977 slightly more than $38 billion in financial aid to less-developed countries. In 1974 the figure was $7.6 billion; in 1975 $11.5 billion; in 1976 $9.0 billion; and in 1977 about $10.0 billion. The figure for 1978 is slightly down from 1977. Controversy surrounds some of these figures. The calculations of the United Nations Conference on Trade and Development (UNCTAD), given above, are larger by about 20 percent than the figures of the Development Assistance Committee of the Organization for Economic Cooperation and Development.

The principal reason for the difference is that OECD does not count as aid the OPEC contribution to the IMF oil facilities since these funds are still reserves on which members may draw. UNCTAD argues that these are new contributions, a new source of financing for less-developed countries, and were contributed specifically for that purpose. They are funds for assistance that would not have been available but for the OPEC contributions. The amount of the contributions for the four years is about $7.0 billion.

Whatever figures one uses, OPEC aid is a large amount, absolutely and relative to total aid. In 1976, for example, the OPEC aid of $11.5 billion, including the $3.3 billion in IMF oil-facility contributions, compares to $17.0 billion in official development assistance and other aid flows of the seventeen most important industrial donors. Although OPEC members are not trying to compete with the United States and other industrial countries in providing aid, they have mounted programs of substantial magnitude.

As befits the largest oil producer, Saudi Arabia has contributed more than one-third of all OPEC aid. In 1976, indeed, Saudi Arabia's aid was second in the world only to that of the United States. Kuwait's contribution of 19 percent of the total OPEC aid outweights Iran's part, only 13 percent, followed closely by the United Arab Emirates with 12 percent. Then it drops down to Venezuela, which contributed 7 percent of the total. The remaining 15 percent is scattered among all the other members. The four Arab Persian Gulf members have contributed two-thirds of the total. Relative to its size and resources, little Qatar is the largest contributor of all.

The ratio of OPEC aid to gross national product is high. Playing with the figures can yield all sorts of percentages, but by all the measures the ratio reflects a large commitment. For 1977 the OECD estimate (lower than that of UNCTAD) indicates that OPEC members' aid was 3 percent of their gross national product, and even the amount of concessional aid was 2 percent. The UNCTAD figures show about 4 and 3 percent respectively. These compare to 0.3 percent of gross national product for the seventeen large industrial countries granting aid.

Data for individual countries display an even more impressive record. For 1976, when all OPEC foreign aid was 4 percent of members' gross national product, Qatar was providing an astounding 19 percent and the United Arab Emirates more than 12 percent of their gross national products. The largest aid-giver, Saudi Arabia, used 9 percent of its gross national product for foreign aid and Kuwait 11 percent. Even poor OPEC countries gave a significant portion of their income. Nigeria, for example, gave more than 1 percent.

Industrial countries fared poorly by comparison. The United States' foreign-aid program was 0.25 percent of its gross national product, Japan's was 0.20 percent; and the Federal Republic of Germany's was 0.31 percent. The Netherlands and Sweden made the greatest relative sacrifices, each providing 0.82 percent of their gross national product in foreign aid. Two OPEC members, Saudi Arabia and the United Arab Emirates, in 1976 ranked among the six largest, in absolute terms, of the bilateral aid-givers.

Foreign aid comes in many different forms. One important distinction is between loans made on regular terms and those which are easier on the borrower. The latter, concessional loans, charges no interest or interest that is substantially below the market rate. Such loans also have a grace period, five years or even more, before the borrower must start paying back. The pay-back period of the loan also is often longer than regular terms.

In OPEC aid more than one-half of the total is concessional. UNCTAD argues that some of the characteristics of OPEC aid—long maturity and grace period, untied loans, its origin in capital not current production—vest it with even greater concessional elements. In addition, if the contributions to the IMF oil facility—which are lent on regular terms—are left out, then three-fourths to four-fifths of OPEC aid is concessional, not much below the concessional part of industrial-country foreign aid.

OPEC foreign aid has now spread to most of the world's less-developed countries. Between 1970 and 1973 only twenty-three countries, mostly Arab and African countries, benefitted. In 1974 the number of countries jumped to forty-two, in 1975 to fifty-five countries, and in 1976, sixty-three countries were recipients. Counting Opec Special Fund operations, eighty countries or more now receive some form of OPEC aid. By 1977, nearly 40 percent of this aid was for non-Arab countries.

OPEC SPECIAL FUND

A sign neatly engraved in gothic letters that is placed on the Palais Strudlehof in Vienna says: "Opec Special Fund." Inside the building a few professionals and secretaries run a $1.6 billion operation that in 1977 reached 67 less-developed countries. Soft of voice, professorial in tone, Dr. Ibrahim F.I. Shihata, one-time professor, runs the show with quiet efficiency.

In 1974 the only certain thing was that there would be OPEC aid programs. Most members already had programs and were busily engaged in expanding them. That was the year that OPEC foreign aid quintupled. What was not certain was whether or not OPEC as an organization, independent of any member, would have a foreign-aid program. The answer did not come until 1976.

OPEC members have a curious attitude toward their organization. They recognize the need for a headquarters and a secretariat. But they do not want to give it any money or power. They fear that a large central organization would soon try to dictate policy and make decisions, detracting from the autonomy and independence of action of each member. A central aid program, they feared, would fortify the headquarters organization, enabling it to allocate funds and influence decisions. When the members did establish the Opec Special Fund, they kept it separate from the OPEC secretariat. The two organizations are not even in the same building.

Many members made proposals for an organization aid program and all through 1974, 1975, and 1976, the members debated. The proposals for a big program, for a joint program with industrial countries, for a program channeling all OPEC members' aid lost out. Members were not willing to turn over to a central program the money and authority they wished to use themselves. In the end, they established in January 1976, a modest, even small program. But they designed it for quick operation and maximum impact.

The Opec Special Fund began operations in August 1976, with $800 million. The first act was to commit more than one-half of it— $435.5 million—to the newly formed International Fund for Agricultural Development (IFAD). Every member gave something, from little Ecuador's $0.25 million to Iran's $125 million. The largest contributors include Iran, Saudi Arabia ($106 million), and Venezuela ($66 million). These three countries are in second, third, and fourth places among the contributors to the $1 billion IFAD fund, behind the United States, but ahead of Japan and Germany. OPEC linked its contribution to a 150 percent match by industrial countries, coercing them into contributing. But for OPEC, IFAD would not exist.

The next step was a program to assist the less-developed countries most seriously affected by what OPEC considered to be the general economic crisis, not an oil crisis alone. The Opec Special Fund made loans to help them pay their bills, but in the context of specific projects. Using a $200 per person income as the base for defining the most impoverished countries, the United Nations counted forty-five such countries. Within months the Opec Special Fund committed $1 million to $22 million to each, the amount depending upon the measures of the degree of need. The loans had no interest, a 0.5 percent service charge, twenty-five-year maturity, and a five-year grace period. In nine months about $200 million more of the Opec Special Fund was gone.

The International Monetary Fund began selling its gold in 1976, making a profit. Most of the members of OPEC, through the Opec Special Fund, contributed their share of the profit, eventually $60 million, to the IMF Trust Fund for further assistance to less-developed countries. Thus, by mid-May 1977, nine months after it began, the Opec Special Fund had spent or commited about $700 million of its original $800 million, plus $60 million in gold profits. In the meantime, in late 1976, OPEC members endowed the Opec Special Fund with another $800 million. By 1979 the commitment of small loans had reached $525 million, and the total Special Fund commitment had reached over $1 billion.

By mid-1978 the $1.6 billion Opec Special Fund still had not used up all its capital. It has continued its concessional lending, mainly to most-seriously-affected countries. It has broadened its lending to include more project loans. Of these, the fund has most often acted as a cofinancier along with other lending agencies on projects already reaching the implementation stage. The fund also made a modest contribution to the United Nations Development Program. By April 1979, two and one-half years after the fund began, only $600 billion of its $1.6 billion capital remained uncommited. OPEC was considering another increase in the fund's capital.

The largest contributors to the Opec Special Fund are Iran and Saudi Arabia. Each provides a little over one-fourth of the total. Venezuela contributes 15 percent and Kuwait 9 percent. The remaining 26 percent comes from the other nine members, all of whom participate to some degree. The fund is actually a series of accounts, one with each member, and a central fund account. When the money is ready for use, the member transfers the appropriate funds to the recipient through the central account of the fund.

The sums handled by the Opec Special Fund have been modest but they have been put to work speedily and effectively. More than one-half of the country loans have been used to pay for capital equipment for industry and nearly one-fourth for agricultural equipment. The loans require that the borrower commit counterpart domestic funds to specific development projects. In this way, the fund spurs projects in less-developed countries. More than one-half of the counterpart funds have been committed by borrowers in power development, with nearly one-third more going to agriculture and agro-industry.

If the funds of the Opec Special Fund seem trivial, bear in mind that they are in addition to the aid that comes directly from OPEC members. Most of the OPEC aid programs remain in the hands of members, implemented by them or, through national and international institutions, selected by them. Thus, there exists a modest OPEC aid program administered by the Opec Special Fund, but the bulk of OPEC foreign aid goes either directly from members to recipients or through other organizations. Even in the Opec Special Fund, the members carefully monitor the use of the money. The Fund Governing Committee and the OPEC Ministerial Committee on Financial and Monetary Matters supervises its activities.

THE MEANING OF OPEC AID

What does OPEC aid mean? What are the implications of aid programs amounting to nearly $10 billion a year carried out by a group of less-developed countries? The easy answer is that OPEC members simply have lots of money, a bountiful financial surplus, some of which they took away from the less-developed countries, and for which they have no good use. It is only right that the less-developed countries receive some of it back.

By that reasoning, the United States would increase its aid whenever the price of coffee goes down or the price of trucks and tractors go up. No industrial country is prepared to tie its aid to the prices of the products it imports or exports. If the industrial countries did, their foreign-aid programs would be several times their present magnitude.

The fact that OPEC even has an aid program is a credit to its members and an incredible development in international relations. The prices of many products have gone up sharply—many did at about the same time as that of oil—but none of the exporting countries, rich or poor, mounted foreign-aid programs as a result. To be sure, OPEC members have political motives. Political and military motives underlie the aid programs of industrial countries as well. OPEC members do not have the additional motive of industrial countries, that of benefits to their own economies.

One fact is certain. The OPEC aid programs shame the industrial countries. They have had decades to build aid programs of less than $20 billion annually. Within a year, OPEC's aid programs were nearly $8 billion and soon were one-half of what the major aid-givers provide. Within a short time, OPEC has become a major element in the financial support of many less-developed countries. The United States and other industrial countries suffer by comparison with OPEC efforts. The OPEC initiative is not lost on the less-developed countries. Their search for exemplars and leadership gravitates toward those who have power and demonstrate responsibility.

Make no mistake, money is power. Even though the OPEC aid programs are smaller, perhaps less certain and less experienced, they are consolidating the influence of OPEC members among less-developed countries and vesting OPEC with clout in the councils of the mighty. Oil may be the root of OPEC wealth, but OPEC's management of its wealth is escalating the international political influence of its money.

The OPEC foreign-aid programs prove that the industrial countries have no monopoly on the ability to mount and maintain large aid programs and manage financial flows. OPEC members have created institutions of worldwide scope, innovations in the aid field. They have initiated and sustained a large transfer of wealth from industrial countries to less-developed countries through themselves. The flow has been directed to countries in the most dire need, with money put to use effectively and quickly.

The probabilities are that the size of the OPEC foreign-aid programs will remain large but will decline somewhat. Many of the aid-granting institutions that channel OPEC aid are capitalized and can continue to provide aid, but not at the levels of recent years. Clearly what OPEC has done and will do relates to the exchange earnings of members and the surpluses resulting from the monetization of the members' depleting assets. As their imports increase members will need their money urgently for domestic economic development, and the surplus of most members will decline, restricting the amount available for aid.

The OPEC aid programs will remain. Indeed, one aspect of aid, direct investment by OPEC members in less-developed countries, has not yet been explored. OPEC members may be able to work out with the less-developed countries more congenial investment arrangements than industrial countries have experienced. This would provide OPEC members with investment returns more nearly competitive with capital markets and would be an additional source of funds for economic development.

Members regard their aid programs as a permanent part of their foreign policy. Some are already borrowing on harder terms than they are lending. Members' policies and actions indicate that they do not feel they can abandon programs that support their new leadership roles, no matter what the cost. The future of the programs and OPEC's relations with less-developed countries depend on the future of OPEC oil. Hundreds of studies of future oil supply and demand conditions have resulted in an equal number of different scenarios, some in sharp disagreement. The next chapter examines some of the forces that will be at work in OPEC and the oil industry in the next generation.

Chapter 10

Peering into Oil's Future

Seeking to divine the future is a hazardous pursuit. In the crystal ball everyone sees the same moving clouds. To some the clouds form a dread figure as when a child sees a giant monster rushing ahead of the thunderstorm. To others, they are the billowy substance of "cloud nine." Most often, people see what they want to see.

The oil companies visualize a great future seeking and finding oil. This vision of the future will involve increasing costs, for which they will need hefty profits. Governments of oil-consuming countries fear a washed-out bridge down the road, requiring them to summon popular support to brake oil consumption now to avoid a calamitous spill later. Most consumers still daydream in happy ignorance. OPEC, with an eye on the rapid exhaustion of their oil reservoirs, foresees the need to design a strategy that will insure its member's economic development before their oil is gone.

Looking at the surprises that the earth and man's ingenuity have already wrought, some researchers see little cause for alarm. The amount of oil is sufficient, they say, to last until the substitutes are ready. Independent researchers, no matter how objective, calculate the future as an extension of the present. They are unable to calculate the twists and turns of trend lines. Still, most of the experts are waving red flags. Only one event do all see, although some see it with clarity, while others see it dimly. At some time in the future oil's price will be astronomically high as oil as an energy source disappears.

FUTURE PHYSICS AND ECONOMICS

The future must obey the same physical laws that the past adhered to and that the present observes. The first and second laws of thermodynamics also demand acceptance of the future. These empirical observations can be bent but not broken. The First Law states that the amount of energy is a constant; energy may neither be created nor destroyed. This seems comforting, however, the law holds for the physical universe, not simply the earth, or North America, or the United States. Nor does it specify the form of energy. A lot of energy exists in the difference between a body of water at $-50°$ celsius and $-60°$ celsius. In neither case, has anyone figured out how to extract and use that energy. What is important is not the total amount of energy—the constant—but rather the energy that can be used to do useful work for human beings. That is where the Second Law of Thermodynamics comes in.

The second law allows any particular form of energy such as oil, coal, and uranium, to exist in limited amounts. It also states that the energy capable of doing work by doing work gradually becomes energy not capable of doing work. Every day as man uses coal or oil to perform tasks, a smaller total stock of all energy remains that can perform work. Far, far out in the future, the same amount of energy will exist, but none of it will be capable of working for man. There's no need to worry; it's so far off that no one need by concerned for many millennia.

Still, there is something worrisome. Man uses the most accessible and cheapest forms of energy first. When he uses that up, he then uses a costlier form. He invents machines to render kinds of energy previously thought to be useless for work into a useful form. He invents machines to convert that energy into work. Both processes require energy and other limited resources. So useful energy costs more and more as man is able to make and use energy and further depletes the limited supplies of useful energy. In a speck of time— say a century or so—this may not be observable, for technology may seem to forestall the inevitable. But the costs for energy that can be used rise inexorably.

Dr. Nicholas Georgescu-Roegen of Vanderbilt University, in his book *The Entropy Law and the Economic Process* (Harvard, 1971), explores the economic implications of the first and second laws. High entropy is what useful energy becomes when it cannot do further work. It is chaos, disorder, wasted heat, uselessness. Entropy increases constantly, just as the energy available for work constantly decreases as man uses it. Try as we may, we can't really beat the entropy law. Technology can only appear to bend it and then only for a moment. The energy-entropy relationship highlights dramatically man's economic short-sightedness and irrationality in refusing to conserve the low-cost useful energy and in using wastefully work-laden energy forms. How much should man borrow from his children by making them pay accelerating energy costs?

Exactly when the earth runs out of oil is not important. Long before that time no one will be willing to pay its price to run a car, heat a house, or power a factory. Some other energy form will be cheaper than oil. What is relevant is the time between now and the first half of the twenty-first century, while oil is still competitive and substitutes are few. During this period, consumers will want to buy oil at different prices, and producers will be willing to sell oil at different prices. From these two relationships the oil market emerges, determining the amount bought and sold and the price.

Most men and nations have long regarded the market as the best way to resolve economic conflicts between buyers and sellers. The oil market of the future may be an exception. OPEC members may suffer if they produce to meet the needs of oil-importing countries. If members try to satisfy the needs of the industrial countries, the importers may suffer. The market will hurt both buyers and sellers because of price and amounts demanded and supplied.

PREDICTING THE FUTURE

Even at its simplest, predicting the future is a perplexing task. Before starting, the researcher must make the agonizing choice of what method to use. Historical trends, changes in trends, functional relations—equations—fixed or changing in time, and specific bits of hard information must be interwoven with the assumptions about how people behave and produce things. The data for each producing and consuming country mix with the method to yield supply and demand estimates for each form of energy. The differences between supply and demand generate trade flows—imports and exports—among the countries.

The future price of oil is always a key assumption. Since price both influences future production and consumption and is determined by the market that sets production and consumption, the modern Merlins must fix the price in order to determine the future market results. Sometimes they use the present price. Frequently they choose the present or a recent real price, that is, the price that represents the purchasing power of a barrel of oil at a specific time. Sometimes they pick a price out of the hat, say $7, or $15, or $30. Sometimes they do their calculations with many prices.

Suppose that a study uses the present real price and determines that oil-importing countries will want to import in 1990 some 53 million barrels a day. This may be greater than the estimated capacity to produce, which is, say, 45 mbd at that time. This is evidence that the present price is too low and will not clear the market. So the researcher tries again, using a higher price. By successive tries he finally balances imports and exports at some price within capacity limitations.

The ardor and uncertainties of predicting the future have not deterred many from making the attempt. Indeed, everyone seems to have an estimate that they revise from time to time. The United States government has produced dozens of estimates, including those of the Federal Energy Administration, the Department of Energy, the Central Intelligence Agency, the Bureau of Mines, other government agencies, and several congressional committees.

The Organization of Economic Cooperation and Development— the organization of the United States and the major developed countries who consume most of the oil—and the International Energy Agency have produced many predictions. Individual European countries as well as Japan have made studies. OPEC also has "guesstimates" about where the oil market is going, and individual oil-exporting countries make independent estimates to know better where they are headed. Because of the interdependence among nations, most studies are global.

The oil companies—Exxon, Shell, Mobil, and others—also have projections into the future. The American Petroleum Institute and its counterpart in other countries study the future oil market. Research institutes, individual oil experts and consultants, and scholars are producing reams of paper and thousands of pages of computer print-outs. All the studies would fill a library. Nearly every year, new predictions or revisions of old predictions come out.

I would be remiss if I did not produce an estimate of future oil supply and demand. It is included in the table below. It differs from the others in that it is not based on independent and original

Table 10-1. World Oil Supply, Demand, Trade, and OPEC Capacity (in mbd)

World Oil Demand	1975	1980	1985	1990
United States	16	20	23	26
Canada	2	2	3	3
European OECD	14	17	20	22
Japan	5	7	9	10
Other developed countries	1	1	2	2
Developing countries	6	7	8	11
OPEC	2	3	4	5
Total demand	46	57	70	79
World Oil Supply and OPEC Capacity				
United States	10	10	11	11
Canada	1	1	1	1
European OECD	1	3	5	5
Japan	0	0	0	0
Other developed countries	2	2	3	3
Developing countries	4	4	6	6
OPEC	28	40	44	48
Total supply	46	57	70	79
OPEC capacity	36	40	44	48
Imports (−) and Exports (+)				
United States	−6	−10	−12	−15
European OECD	−13	−14	−15	−17
OPEC	+26	+34	+40	+48
Rest of the world (net)	−7	−10	−13	−16

research, but rather, is a gross average of the estimates of others. The only claims for it are that the arithmetic is right and that it is as reasonable as anybody else's guess. And like them, it is undoubtedly wrong.

Table 10-1 centers on the oil balance in 1985 and 1990, but many estimates from other sources also exist for 1980, 1995, 2000, and later dates. The usual form of estimates is a table, usually more elaborate than the one in this chapter. This table shows production, consumption, imports, and exports for the United States, OECD,

OPEC, and the rest of the world. Often other energy forms are included. In my table, supply always equals demand—what the world imports, the world also exports. OPEC exports, indeed, are calculated as a residual, as befitting its present position as the residual supplier. Estimated supplies outside OPEC are subtracted from total supplies to equal OPEC exports. Net import deficits for the rest of the world must equal the OPEC export surplus.

OPEC production and exports may not agree with the figures in my table. The figures say, rather, that in order to balance the world's oil accounts, OPEC must produce and export those amounts. This begs the question of whether or not OPEC can or will produce enough to balance the accounts. Part of the answer to this question lies in comparing the OPEC production estimate with an independent estimate of OPEC's capacity to produce. This estimate does not depend on the supply and demand estimates. If capacity exceeds the production required to balance the accounts, then no problem necessarily arises. Whether or not OPEC will produce that amount remains an open question.

DEMAND

How much oil will the world want to buy in 1985? How much of it must come from OPEC? At what price? *Oil Daily* in early 1978 reported that Arnold Safer of Irving Trust of New York was predicting a decline in OPEC exports and prices by 1985. Secretary of Energy James Schlesinger said in mid-1978 that the world will need 37 to 39 mbd from OPEC and will have to pay a substantially higher price than they now pay. The estimate of the CIA is even higher than that of Dr. Schlesinger, showing Soviet net imports and the demand for Saudi Arabian oil three times the present demand. The OECD estimate, the calculations of the Brookhaven National Laboratories, the Exxon prediction, and the IEA estimates are in the same ballpark as that of the Department of Energy. All predict higher prices.

The demand for oil hinges on many factors. They include the responsiveness of demand to price, that is, the reaction of consumers, the level of income in the future—economic growth between now and the future date—the oil-output ratio—the amount of oil used per dollar of gross national product, the rate of introduction of substitutes, such as coal and nuclear energy and their prices, and energy policies. The demand for oil embraces the whole panorama of production and consumption relationships in the economy.

The demand for OPEC oil depends on the production and consumption of oil in the oil-importing countries, the exports of other oil-exporting countries, and energy policies. All of these are interrelated and each in turn depends on other economic, political, and physical considerations. About some determinants of demand there is little disagreement, but about others, such as economic growth and the oil-output ratio, wide divergences exist.

Most studies agree that demand is unresponsive to price. In a very short period the same amount of oil will be purchased regardless of price. Over longer periods, demand may be more elastic as consumers decide to buy less at higher prices, doing without or using a substitute. Time is required for people to change their minds and their buying and living habits, and for industries to change production practices. Greater responsiveness of demand to price would, or course, have the effect of reducing demand more as the price increases.

The probable case is that demand responds only slightly to price changes in the low price range. Changes in the range of $5 to $15 elicit only a minor change in the amount people want to buy. Changes in the range from $15 to $25 may spur more action by consumers. Changes in the range from $25 to $40 may begin to hit hard at demand. The reason, of course, is that at higher prices, the impact on the user's budget is greater. He feels it more. In addition, in the higher price ranges, substitutes may become cheaper than oil. These price effects work only with real prices, prices with the inflationary element common to all prices removed.

Changes in nominal prices sometimes fool buyers, but over a period of time, it is the real price that counts. An enterprise paying 22 percent of its costs for oil will not buy less at the higher price unless its real costs go up. If oil becomes 28 percent of costs, then the firm will cast about for ways to reduce oil costs. An absolute price of $35 a barrel in 1985 or $57 a barrel in 1990 may have little influence on demand if inflation proceeds at 10 percent per year. What incentive does the motorist have to cut gasoline purchases if his housing, food, and clothing costs, along with his income, go up as fast or faster than the price of gasoline?

Taxes bear heavily on demad. The price of a barrel of oil is not what the consumer pays; he pays much more. The cost of crude oil is one-third to one-half of the price of oil products. The rest of the cost of oil products consists of payments to the company for its various operations and to the government in taxes, which in Europe are nearly one-half the price. If the price of crude oil goes up and taxes go down, with oil-product prices staying the same, the consumer will not respond by economizing.

DEMAND AND ECONOMIC GROWTH

Increases in the level of income and production—economic growth—pushes oil consumption up. Behavior and technology tie oil to how much people earn and spend and how much industry, agriculture, and commerce produce. With a larger income the consumer must heat his larger home and power his bigger car on longer vacations. With greater production in the economy, enterprises must use more oil in more factories, stores, and farms.

If the economy fails to grow, oil consumption may change only slightly. A recession—a loss of real production and income—even reduces the demand for oil, as it did on a worldwide scale in 1975. Arnold Safer got his low demand estimates for the 1980s by assuming a major recession in 1981 and 1982. But predicting the rates of growth of the American economy, much less all the world's economies, is a risky business.

Economists have enough trouble estimating for the next quarter or next year gross national product in the United States, the paradise of economic data. They have intricate mathematical models that try to take into account almost every influence. Their figures are nearly always off, sometimes by wide margins. The experts are trying to predict what millions of people and thousands of enterprises will decide to do even when these people and enterprises have no idea what they will do until the time comes.

Let's suppose that reliable estimates of future income and production exist. The link between them and the demand for oil must then be known. This link is the oil-output ratio. Prices, habits, production processes, technology, lifestyles, stocks of oil-using capital and consumer goods, price and availability of oil substitutes, energy policies, and many other factors define the link. The ratio changes as time passes and is different in different places. In the United States, for example, the ratio is twice that of many other oil-importing countries. The oil-output ratio tends to move sluggishly and only when everything pushes it in a single direction.

Some believe that the oil-output ratio of the United States has already begun to fall under the impact of the changes since 1973. If it has, it can effect large changes in the demand for oil. Halving the ratio, which would represent a heroic change in the economic, political, and social ways of doing things, would reduce the historical rate of increase of oil consumption to 3.5 percent. This rate is close to the rates employed in many studies of future demand. A further reduction in oil consumption's rate of increase would result from the future economic growth of the economies of oil-importing countries being slower than in the past.

One important reason for the decline in the oil-output ratio might be the increasing use of substitutes. Coal, nuclear power, and other substitutes can replace oil in many uses now. At some point, the price of oil and of the substitutes will force utilities, homeowners, and consumers to switch. The introduction of substitutes, however, takes time. It takes at least ten years to get a nuclear power plant on line. Changes in coal-using facilities takes years. Seven years are required to get a new coal mine in operation. In addition, the price advantage of the substitute must be permanent and decisive in order to overcome not only the cost of the substitute but also the cost of replacing oil-using equipment with equipment for the substitute. Substitutes will make progress, but experts do not believe that their contribution will be great before 1990 or the turn of the century.

Some experts predict slower economic growth in the future, giving oil as the reason. Whether or not a country's economy grows depends on all factors of production—labor, resources, capital, and entrepreneurship—as well as conditions that determine their productivity, such as technology, incentives, the socio-cultural matrix, and many other elements including government policies. Oil is only one small ingredient in the growth process. Countries will grow or not grow at one rate or another regardless of oil.

It must be said that the knowledge on which to base judgments about both the rate of growth of economies and changes in the oil-output ratio is extremely limited and fragile. Despite vast amounts of economic information, tomorrow's decisions follow tomorrow's muse, who is unknown today. Often researchers come dangerously close to assuming the results of their research by postulating rates of change and coefficients that make the conclusions ineluctable. When one reads press reports of one study or another, it is easy to accept the expert's words. After all, they have studied the matter and should know. The press seldom prints the assumptions and biasses or the fine print of the study. So many contradictory or, at least, widely divergent studies have now been reported that each new study should be greeted with caution.

Some studies simply finesse the task of working out economic growth rates, oil-output ratios, and other technical matters by estimating the rate of increase of oil consumption directly. The rate is often based on the presumed historical rate of nearly 7 percent, which in turn is based on historical growth rates of about 4 percent. Some studies are also influenced by the rate of increase of world exports, which was 11 percent per year from 1960 to 1973.

Nearly all the studies foresee a substantial decline in consumption's rate of increase. Falling growth rates of the world's economies as well as reduced oil-output ratios underlie these predictions. The resulting rate of growth of the demand for crude oil is usually 2 to 4 percent, quite modest in view of past experience. If the world's economies boom in the 1980s, however, and the oil-output ratios decline only slightly, world crude demand could increase by as much as 6 percent per year.

The power of compound interest is great. At 3 percent, any starting figure doubles in twenty-four years. At 6 percent, it quadruples in the same time. The 16 mbd demand of the United States in 1975 (see table) becomes more than 21 mbd in 1985, and becomes 25 mbd in 1990 at 3 percent. Increasing at 6 percent, the 1985 demand becomes nearly 29 mbd and more than 38 mbd in 1990, 2.5 times greater than calculations using 3 percent.

The oil and energy policies of oil-importing countries underlie their future demand for oil. By taxation of oil products and equipment using oil products, expenditures on substitutes, regulations, and direct controls, governments can either encourage or discourage the use and production of oil. In many countries, governments mandate the use of one fuel or another in equipment and regulate the efficiency of oil-using equipment. But other policies, such as the United States', control the price of oil products below the market price, in effect encouraging the use of those products.

Most countries have already adopted moderately stern oil conservation policies. Often, as in most of Europe, oil conservation is achieved only by using more natural gas. So far, only the United States acts almost as though nothing had happened. The United States' 1978 Energy Act is some improvement in recognition even though the legislative process pulled most of the teeth from President Carter's proposals. Oil profligacy in the United States in the 1970s, however, could represent a potential reduction in oil demand in the 1980s should the United States decide to employ stronger measures.

The demand for OPEC oil exports is not the simple sum of all oil demand. Many of the large oil-consuming countries are also oil producers, and some oil producers are becoming important oil consumers. The demand for OPEC exports depends on the supplies of non-OPEC oil and the OPEC demand for oil. If the oil-importing countries, particularly the United States and some of the less-developed countries, can boost their own production enough, they could ease the pressure on OPEC production.

Despite variations in estimates, most studies center around a world demand for oil of about 57 mbd in 1980, 70 in 1985, and nearly 80 mbd in 1990. The World Energy Conference predicts 60 mbd for 1980 and 89 mbd in 2000. Each study is different, and the figures in the table included in this chapter are a gross average. The figures exclude Communist countries, which in most studies will neither export nor import in large amounts. CIA reports that the Soviet Union will soon be a large importer. Most exporters don't think so. China could be an exporter, but no one knows for certain.

The two largest components of world demand are the United States and the European OECD countries. Japan and the other developed countries follow. U.S. demand in 1975 was 35 percent of the total; European OECD was 30 percent. Other developed countries took 17 percent of the total, and developing countries, including OPEC, took another 17 percent. In the future the U.S. share will probably remain about the same, but European demand will not increase as rapidly as that of the United States. Japanese demand will increase strongly because of its almost total lack of oil.

The rate of increase of demand from 1975 to 1980 for most studies is more than 4 percent. But 1975 was a depressed year, feeling the effects of the quadrupled price, the embargo, and the recession. In the table in this chapter, the 1975-1980 demand increases 4.4 percent. From 1980 to 1985, the rate of increase declines to about 4.2 percent, and from 1985 to 1990 it declines to about 2.4 percent. The overall increase from 1980 to 1990 is about 3.3 percent. These rates of increase imply from past experience a large reduction in all aspects of oil's use, representing a cut of nearly 50 percent in historical rates of increase.

SUPPLY

Is the supply of oil going to be sufficient to meet the forthcoming demand? No. supply will fall short at present prices and market arrangements. But at the prices and methods for allocating oil at future dates, of course supply will meet demand. Through the market or by coercion, supply will always equal demand. If people want to consume more than suppliers will provide, then the price will rise, squeezing buyers out of the market or, by some form of controls, buyers will be removed from the market and more supplies may be brought in. Many problems will arise in allocating supplies, but consumers cannot buy more oil than producers sell.

What determines supply? The quantity of oil in the earth, the capability of extracting oil at any given moment, the price at which oil sells, and the costs to produce it all help to determine oil supplies. Underlying these factors are technology, the organization of the industries, and the prices of labor, capital, and talent required in exploration and production. And overriding all the factors is the willingness of producers to sell their oil. Bear in mind that oil, to the producers, is a nonrenewable fixed asset. The producers' subjective assessment of the value of the asset over a long period of time heavily conditions their supply price.

The absolute amount of oil in the earth is not relevant for the next generation or more. Oil in the deep sea, under the polar caps, and in inaccessible regions may become important in the next century. But if some of oil's existence is completely unknown or unsuspected now, it will not have much of an effect on supplies for the next quarter century. Even so, some of the supplies for 1985 and 1990 must come from probable rather than proven reserves.

The absolute amount of oil is relevant for a general picture of long-run supplies. Exxon, among others, reports that about one-half of the world's oil has already been discovered and has either been used or is now in proven reserves. Most estimates place world proven reserves at the beginning of 1978 at about 650 billion barrels and consumption from 1859 to 1978 at about 370 billion. By the Exxon estimate the total stock would be about 2 trillion barrels, of which 20 percent has been used and another one-third more has been discovered. Future additions to reserves, according to estimate, will be about 40 mbd (about 15 billion barrels) each year in the coming years, but varying greatly year to year. As Exxon puts it, "The normal draw-down of reserves will exceed the average rate of new discoveries by steadily increasing amounts in the future" (*World Energy Outlook*, Exxon, 1978, p. 30).

Still, in order to balance supply and demand in 1990, the world must rely upon oil not yet discovered. Nearly one-fourth of U.S. consumption in 1990 must come from additional oil believed to exist on the mainland, in Alaska, and in the offshore Atlantic. Nearly one-third of world consumption before 1990 must come from probable reserves, both inside and outside OPEC. The existence of these reserves is nearly certain, but the detailed exploration has not "proven" them so far. No configuration of proven reserves, probable reserves, or new discoveries outside of OPEC, however, will eliminate the heavy dependence of oil-importing countries on imports from OPEC, before or after 1990. Its share of proven reserves now is more than two-thirds of the world total, and it is unlikely to change in this century.

OPEC has about 450 billion barrels of proven reserves, enough to last about 40 years if used at production levels of 1977 and 1978. Adding some for new discoveries and improvements in the recovery rate, subtracting OPEC's increasing consumption and present export levels, the oil-exporting countries will face declining production shortly after the turn of the century and the oil will be gone during the first half of the next century. The lack of oil supplies in the oil-importing countries implies that present export levels are insufficient; these countries want increasing oil from OPEC members. Used without constraint, OPEC oil could disappear by the turn of the century.

CAPACITY

Capacity is the more immediate restraint on supply. There are several definitions of capacity. Installed capacity represents proven reserves, wells, and design limits on all engineering equipment and facilities in production, processing, transportation, and storage of crude oil. It is the maximum limit of production for a moment with everything going full tilt. It is never reached.

A more reasonable estimate is sustainable capacity. It is the production rate that can be maintained over a period of time, usually three or four months. Production seldom reaches ninety-day sustainable capacity. Sustainable capacity cannot be maintained over a longer period of time without continuous new additions to proven reserves, new wells, new equipment and facilities, and maintenance.

Still another capacity concept reflects the policies of the producer. It is the available capacity, the amount the producing countries believe should be the maximum production rate in view of investment plans, market conditions, conservation of the fields, and oilfield conditions. In the late 1970s it also reflects the voluntary restraint of OPEC members to maintain the price. In 1978 available capacity is estimated at 33 mbd, reflecting excess capacity of no more than 10 percent, since OPEC production is more than 31 mbd. Some of this excess capacity is necessary to accommodate seasonal changes. Installed capacity, the maximum and in fact unattainable capacity, is almost 41 mbd. The *International Energy*

Statistical Review (June 14, 1978, p. 3) estimated in mid-1978 that estimated 90-day sustainable capacity was less than 37 mbd.

As more facilities are built and more wells brought in, the amount of capacity increases. For 1980 and other future dates, the relevant supplies available are not the needs of oil-importing countries, but rather the capacities of OPEC members. Many estimates have been made of future OPEC capacity. Great uncertainty clouds these figures. Physical limitations of the oilfields intrude. In some instances, no matter what facilities exist, the oil just cannot be drawn out any faster. In addition, building additional capacity is a deliberate and costly investment decision. Without the intention to produce at higher levels, OPEC members will not build additional capacity, since excess capacity tends to undermine price.

In 1975 OPEC members had about one-third excess capacity. Only heroic restraint prevented a fall in the price. As time passes, most experts feel that OPEC capacity will grow more slowly than demand. In 1980 excess capacity will be no more than 8 percent, not enough to put any downward pressure on price. The 1980 capacity of 40 mbd represents an expansion of 8 percent in capacity from mid-1978. Indeed, if OPEC fails to build more capacity, there will be no excess capacity in 1980—a signal for rising prices.

By 1985 no excess capacity will exist even if OPEC expands capacity by another 20 percent. This is a clear case for additional rapidly rising prices. According to many estimates, the tight market of 1979 will merge into a continuing tight market in the early 1980s and by 1985 OPEC will barely be able to meet the demands of oil-importing countries, even with rising prices. The U.S. Department of Energy, before the Iranian revolution, places the oil crunch between 1980 and 1985. The shortfall of Iran's exports, made up for only in part by other OPEC production, could produce upward pressure beginning in 1979 and continuing throughout the 1980s.

Capacity is sensitive to some degree to price and expected prices. Building additional capacity is an investment decision, but it is in the hands of the governments of OPEC members, not private companies. If the price of oil is high and rising, or is expected to rise, the private investor may be induced to add to proven reserves and to construct the physical facilities that permit greater production. OPEC members, considering the unresponsiveness of demand to price increases, may decide not to create new capacity. Production would edge toward capacity and the price would rise, yielding greater OPEC revenues. From the point of view of producing nations, adding to capacity could restrain price increases by increasing production that uses up oil reserves.

Let's suppose that the inflation rate for the next decade or more will be about 10 percent. The world index of export prices increased 159 percent from the third quarter of 1973 to the end of 1977, with the index almost stationery from the beginning of 1975 to mid-1976. Starting with a crude-oil price of $20 a barrel in mid-1979, and increasing at 10 percent per year, the price in 1985 will be nearly $36 a barrel. Since excess capacity will be nearly zero most of this time, OPEC members will probably insist on an increase in real price. If they ask for a real price increase of 5 percent per year, then the 1985 price will be $46 a barrel.

If OPEC produces 44 mbd in 1985, uses 4 domestically, and sells 40 mbd at $46 a barrel, members' earnings will be more than $650 billion, more than four and one-half times present earnings. Some members—Saudi Arabia and Persian Gulf members—are still accumulating surpluses at present prices and volumes. Even with present earnings, many members also recognize that much of the capital they are trying to absorb now is not always effectively used and should be cut back. OPEC members by 1985 would have to quintuple their imports in just a few years in order to use as much as they will earn. This increase is possible but not probable.

Whether or not OPEC changes its present residual-supplier policy, let's suppose that the price in 1985 is $46 a barrel and OPEC exports 40 mbd. What about the other side of the coin—the buyers? The less-developed countries, for example, importing 8 mbd, will pay OPEC members, or someone, over $135 billion for oil. The fact that other less-developed countries, oil exporters not members of OPEC, will earn some revenues from the oil exports, is small comfort to those who must buy oil. It will be a crushing financial burden that could slow or stop development in those countries.

The economic picture isn't bright for the industrial countries either. Japan, for example, will pay out $150 billion, the United States nearly $200 billion, and the European OECD countries more than $250 billion. Such huge expenditures cannot help but adversely affect the growth of their economies. This slower growth, of course, will then dim the export prospects of the less-developed countries, further damaging their economies.

The crystal ball becomes very murky after 1985. OPEC would have to add another 20 percent to 1985 capcity by 1990 in order to maintain the market in the same state of tension as in 1985. The experts are divided. Some say that OPEC, by straining, could build enough capacity and produce enough to meet the expected 1990 demand. Others argue that even if it could, OPEC has little incentive to increase capacity and would not be likely to do so. Still others believe that physical conditions limit expansion in OPEC oilfields. Their long-run capacity can expand only slightly beyond 45 mbd, just a little more than 1985 capacity and production. The issue cannot be settled with present information. Certainly, if OPEC is able to meet 1990 demand, it will just barely be able to do so and only with strenuous effort.

The many estimates of supply, demand, and capacity vary greatly. Some anticipate problems arising as early as 1979 or 1980, with production pressing capacity and rising prices. Prices then escalate sharply throughout the decade. By 1990 only stern conservation, large-scale use of oil substitutes, and an all-out production effort by OPEC could balance oil supply and demand. Others see the same process occurring, but beginning in 1985 or 1990, or even later, and think that the crunch can be spread out over a decade or two. The 1978 World Energy Conference set the date at 1990, the Rockefeller Foundation in the late 1980s, and the Workshop on Alternative Energy Sources between 1985 and 1995.

The scenario is always the same. Only the date, duration, and degree of price rise change. The date, of course, is sensitive to price, which acts both as a production stimulant and a conservation measure. It also stimulates the use of substitutes. Ideally, the rising price determines the behavior of consumers and producers, as in any market. The volume of imports and exports of crude oil, however, is so large, that changes in price determine much of the flow of funds in international finance and the economic well-being of both buyers and sellers.

OPEC'S WILLINGNESS TO PRODUCE

Suppose that in 1985 and 1990 OPEC members were able to meet the demand of the world for oil. What does this imply for production levels for each member? The different countries face disparate circumstances. Some, such as Saudi Arabia, can expand more rapidly than can others—Iran, for example. If OPEC meets the 1985 demand, Saudi Arabian production would have to double its present production to nearly 16 mbd. Iraq, Kuwait, and the United Arab Emirates must also increase production. Total exports would have to go up 57 percent compared to 1975.

Now, go on to 1990. OPEC exports would have to climb another 20 percent, 85 percent over 1975. Saudi Arabian production would be about 20 mbd, about 2.5 times greater than in 1978. The Saudi Arabian share in OPEC would increase to more than one-third, compared to the present one-fourth. Iran, for want of crude reserves, could not keep up and production might even be declining further by that time. Indonesia may be importing oil, and expansion in some OPEC members will be limited. For other OPEC producers expansion is possible. In 1985 OPEC members would have to produce as much as they could. In 1990 they would have to produce more than any experts think they can.

The key to future supplies is capacity and production in Saudi Arabia. The other twelve members cannot have the capacity to make large additional contributions to satisfy increasing demand. Saudi Arabia might be able to. Some estimates indicate that the requirement on Saudi Arabia, 20 mbd, is too low. But let's say that in 1990 Saudi Arabia is doing its part in meeting world demand by producing 20 mbd.

Suppose also that Saudi Arabian reserves have remained constant. That is, each year the amount of newly discovered and proven reserves equals the amount removed from the ground. This is a highly optimistic assumption. Under these circumstances, however, how much oil would Saudi Arabia have in the ground if no more reserves are found after 1990? Producing at 20 mbd, they would have enough to last about 15 more years. Make the same assumption about all of OPEC: meeting world oil demand, with sustained reserves until 1990, then no more discoveries. How much would be left? Producing at 53 mbd, they would have enough to last only a little more than 20 years.

Now put yourself in the position of Saudi Arabia. You are earning about $50 billion a year now and can spend only $20 on goods and services. Assuming no inflation—highly improbable—and a doubling of the real price—quite a conservative estimate—Saudi Arabia would be earning about $300 billion a year in 1990. What can Saudi Arabia do with all that money? The economy could not absorb anything like that amount in imports.

The only course of action would be to invest in other countries, primarily the industrial countries. If inflation did exist, and its earnings were correspondingly higher, it would eat up most of the earnings of its huge foreign investments and undermine the value of the assets themselves. Depending on where Saudi Arabia invested, currency depreciation could also threaten asset value. Reasoning from a Saudi Arabian point of view, it would make better sense to keep the oil in the ground rather than risk it in other kinds of assets.

The Midas touch blesses and curses Saudi Arabia. If, as demand rises, it cannot or will not sell sufficient oil to close the gap, the price will shoot up to very high levels and the country will be flooded with money. If Saudi Arabia does supply sufficient oil, the price will rise substantially anyway because of the tightness in the market. The combination of rising prices and increasing volume would still deluge the treasury.

With limited ability to use the money effectively, Saudi Arabia is fated to acquire large financial assets or holdings in other countries. The desert kingdom will be an important but troubled financial power in the world whether it likes it or not. Saudi Arabia will find, as Midas did, that its magic brings long-run disadvantages. Having more of its total assets as real assets—oil in the ground—has decisive advantages. Why should Saudi Arabia violate its own national interests to satisfy world oil demand?

What is true of Saudi Arabia is true to a lesser extent of several other OPEC members, including Kuwait, the United Arab Emirates, and, depending on the price of oil and new discoveries, several other members. Some, of course, could handily use the additional revenues productively in developing their own economies. For those incapable of absorbing capital rapidly and insuring their own long-run development prospects, meeting world demand would mean fast exhaustion of their oil in return for financial assets. The experience that OPEC members have had so far with financial assets does not encourage them to hold more.

If OPEC members produce to meet 1985 demand, they will then have only 330 billion barrels left in the ground. If they continue to produce 44 mbd each year after that, with no additions to reserves, they will exhaust their fields in about twenty years. OPEC would use up another 90 billion barrels of its reserves between 1985 and 1990 meeting 1990 demand. The amount left in 1990 would be about 240 billion barrels, enough for only twelve years. Between now and 1990, of course, more reserves will be proved and improvements in the recovery rate may add to reserves. But capacity and production would begin to decline long before exhaustion. OPEC sees the bottom of the barrel.

OPEC intentions and willingness to supply oil, regardless of price, becomes critical sometime after the early 1980s. Up to that point it will probably be to members' advantage to continue to supply world oil demand as it is expressed in the market place. After that time, Saudi Arabia and others will no longer find it advantageous to accept the dictates of the market. Pursuit of their own interests mandates the refusal to build capacity and the desire to hold back supplies.

The next stage of development of OPEC is how the members will cope with future production levels as well as price policy. The serious thinking has already begun. Secretary General Ali M. Jaidah alludes to it in speaking of the development of a long-run price and production strategy, a process that began in Taif in May 1978 (Interview, *Worldview*, March 1979, p. 42). With Sheikh Yamani of Saudi Arabia as chairman of the group of oil ministers who will study the problem and develop the strategy, the OPEC members expect policies that will eliminate their role as residual suppliers and will change the character and behavior of the oil market.

THE MOVABLE DATE

Members of OPEC will probably in the next few years begin to set maximum production figures, targets beyond which they will refuse to produce. These figures, unlike the ones OPEC members have set up to now to buttress the price in the face of excess capacity and flagging demand, will serve to preserve their real assets. How they will determine these figures and how flexible the production levels will be, no one knows. Under present circumstances, the OPEC moves will be unilateral.

Saudi Arabia's Supreme Petroleum Committee has adopted a plan to increase that country's capacity to 14 mbd by 1984. This is well below the amount needed from Saudi Arabia to satisfy estimated world demand that year. Some experts have estimated that on a continuous and long-term basis, Saudi Arabia, in view of its reserves and the conditions of its known fields, should not produce much more than 12 mbd. These considerations, along with the cutbacks and the reluctance of other members, such as Kuwait and Venezuela, and perhaps the inability of Iran and Indonesia, to expand, suggest that problems could arise even before 1985.

But examine the table in this chapter again. All sorts of possibilities exist for changes in supply and demand that could shift the date of the oil crunch. Suppose that Alaska proves more productive than expected, that the U.S. finds large reserves in the Atlantic, that Mexico becomes a large producer, that China or the Soviet Union, assumed to be just self-sufficient in the table, become exporters. Suppose that technology jells and that new methods are developed to exploit the Venezuelan tar belt or Canadian tar sands. Suppose that a technological breakthrough comes along for some substitute. Suppose major oil consumers embarked on a crash nuclear or coal program.

Many of these events would require more time than is available between now and 1900 or 2000 to lighten the burden in the oil market. The better prospect is for conservation. Slow growth in the industrial countries could restrain the growth of their demand for oil. With sufficiently high product prices, consumers could move the oil-output ratio down. A combination of enlarged production of oil and substitutes and conservation of oil could keep demand within OPEC capacity limits. Serious efforts to resolve domestic energy problems by industrial countries could induce OPEC members to set higher production levels, in the belief that peak demand would be temporary and would recede once the conservation and production programs take effect.

At the extreme, the optimists see that reserves, capacity, and production will be greater than expected and that demand will move more slowly, pushing the crunch off into the twenty-first century. By that time the substitutes, they hope, will be ready and will, in an orderly transition, replace oil. Sometime late in the century, the exhaustion of crude oil as a fuel will coincide with the full takeover by substitutes. No crisis will arise, but rather the economies of industrial countries will adjust gradually without ill effects. This is possible. But frail mortals would do well to keep in mind Murphy's Law: if anything can go wrong, it will.

The critical figures are OPEC production levels of 45 to 55 mbd. Beyond this range or perhaps even less than the lower figure, OPEC will cease to be the residual supplier. Even though the members, especially Saudi Arabia, might be able to produce more, they will not do so on grounds of national interest. Greater production would exhaust their supplies too rapidly, imperil their own long-run development programs, and encumber them with assets of dubious value for which they have no use. No matter how one juggles the figures in the table or what rates or coefficients one uses, it is almost inevitable that OPEC will reach the point of no more expansion between 1985 and 1995.

The possibility exists for persuading the members to opt for the highest figures. Cooperation between the industrial countries and members of OPEC could assure members that their earnings and the value of their financial assets would not deteriorate. Such guarantees, coupled with serious conservation and non-OPEC production efforts, could suggest to OPEC that its interests could be served with higher production.

The United States and other oil importers would have to make a deal with OPEC members. OPEC would be in the superior bargaining position. So far its stance indicates that cooperation must go far beyond oil and embrace all the problems of development, not only of OPEC members, but also of all less-developed countries. The preliminary bargaining in Paris in 1976-1977, which ended in a stand-off, shows that the oil-importing countries so far do not regard their position as perilous and that OPEC will not yet abandon its championing of world economic reform.

The urgency to conserve oil and move to substitutes is not yet observable in the behavior of the oil-importing countries. Most of the industrial countries are indeed trying to conserve and are seeking substitutes, but not with great passion. The United States continues to dally. The soft market of the late 1970s took the edge off the need for immediate action and concealed the gathering storms that lie ahead.

The tightening of the oil market in 1979 was a reminder that oil has the world on a short leash. Fortuitous circumstances, such as the Iranian revolution, oil-company inventory mismanagement, adverse weather, unwise government policies, and many others, can turn a glut into shortages in a short time. The long-term position of oil permits few miscalculations or errors. The danger of the 1979 bouyant market is that it could merge into the chronic shortage that most experts expect to appear in the 1980s. No longer can large users of oil relax when price is stable and the market seems calm. It may be the eye of the hurricane.

THE SCRAMBLE FOR OIL

The scramble for oil begins when the oil-importing countries recognize that OPEC can't or won't satisfy their needs and that no substitutes are ready. Each country or group of countries will seek to get enough oil for its own growing requirements. The competition will be either benign or malignant, depending on the advanced preparations of industrial countries and their ability and willingness to compromise and cooperate on a world scale or to risk and endure conflict.

The world economic and political system is premised on the acceptance of market solutions to most economic problems. The less-developed countries are challenging this proposition, although unsuccessfully so far. The industrial countries cling tenaciously to the method that has benefited them so much and which they believe confers the greatest benefits to all. It is ironic that the next step, for oil markets at least, may witness the rejection of the market solution because it may undermine the continued progress of its participants.

The acceptance of the market as the principal economic guide is based on supply and demand conditions for goods and resources that confer more net economic benefits than any other method. But the world has never before confronted a situation in which a resource built into the very heart of modern industrial economies was in such chronic short supply that the market solution inevitably damages both buyers and sellers. As production presses on limited supplies, not temporarily as in the early 1970s, but chronically as early as 1980 and as prices skyrocket, the problems of the buyers magnify and the market also produces a solution unacceptable to the sellers.

At very high prices some buyers must drop out of the market. The first to default will be the less-developed countries without oil. At yet higher prices, the less affluent developed countries also can't pay. Eventually the price goes so high that it creates serious balance of payments problems for even the richest industrial countries—the United States, Germany, and Japan. All countries feel that they require the oil and will continue to require it. All have economic, industrial, and military needs that make acquiring oil imperative. The substitutes aren't ready and won't be. The market, of course, will balance demand to supply, but the solution—the price and the flow of oil—will cause such extreme hardship for buyers that it will not be acceptable.

Sellers will also drop out of the market. Normally, a higher price would draw in more supplies. But to meet the demand requires that OPEC suppliers cough up their oil at such a rapid rate that they will exhaust their fields too quickly. The development problem is a long-term one. Just to get a firm grip on the bottom of the ladder of economic progress and take a few steps up is a problem that extends for decades. Yet the market says, sell all your oil now. If exporters follow their self-interest, they won't do it.

Their compensation for meeting the demand is the accumulation of financial assets. Even modest inflation, however, will eat away at the value of these assets and eliminate their earnings. When Saudi Arabia or some other exporter needs real resources say thirty years from now, the dollars they earned and salted away in 1979 will be pretty sickly things. The market solution is no more acceptable to oil exporters than it is to those who are economically devastated trying to buy the oil.

The economist says that there must be some price that is satisfactory to both buyer and seller and clears the market. This is true only so long as the financial flows generated by the market do not create problems that undermine the economic health of those that are in the market. With buyers of varying capacity to pay the price and sellers of varying capacity to use the funds, the market, although continuing technically to function, breaks down the framework within which it operates. Buyers and sellers reject the market and seek others means—cooperation or conflict—to allocate the oil and determine its value.

OPEC members feel that to supply the oil needs of the industrial countries will damage their interests. It does not matter whether or not this truly would be the case, so long as OPEC members act on their belief. The industrial countries feel that not to have the oil would stunt their growth, impair living standards, and strangle them economically. Again, no matter whether or not these results will follow, the industrial countries' behavior reflects their conviction. Neither side will feel that the market solution—an astronomical price and modestly rising yet inadequate supplies—is consistent with its welfare.

The storm warnings of conflict are already flying. In a prescient study sponsored by the Rockefeller Foundation, published in 1978 as a *Working Paper on International Energy Supply*, experts from many nations evaluated the potential for conflict arising out of the scramble for oil. Arguing that "no mechanisms are presently available to deal with a situation of chronic tight supply which will be accompanied by rapidly rising prices," the experts warn of "corrosive competition" that "would result also in an erosion of security and of the international economic and financial system." The blandness of the words should blind no one to the scary possibilities.

One or more industrial countries could try to take the oil away from OPEC members. Kissinger once spoke ominously of a military solution if the United States faced economic strangulation for want of oil. In the 1973 crisis the United States made threatening noises. Industrial countries could also try to foment revolutions, hoping that the next regime would be more amenable. They could invade and seize the oil. The risk, however, is great. OPEC members have already announced that they will fight and if necessary destroy their fields rather than let them fall in the hands of outsiders. Even a simple surgical strike could degenerate into an ugly war. The revolutionary trend seems to run against the industrial countries, rather than in their favor.

Alternatively, an industrial country could strike up a special relation with an OPEC country, trying to guarantee its supplies. For OPEC this amounts to joining the opposition. Even non-OPEC members such as Mexico show no inclination for a special relationship. Some evidence of efforts, however, exist in the special relationship that the United States tried to establish with Saudi Arabia before it was soured by the Egyptian-Israeli peace treaty. Revolutionary Iran has also spurned special relations with industrial countries.

The industrial countries could easily fall to squabbling among themselves over the amount of oil available, wrecking international trade and finance in the process. Already in the tight market of 1979 European countries are bitterly critical of the high rate of U.S. consumption. Ordinarily, of course, the price separates the men from the boys. But it is difficult to imagine a situation in which only the few most powerful countries would walk off with all the oil only because they only could afford it, leaving little or no oil for the weaker industrial and the less-developed countries. Nations might bargain, bicker, or even fight to get the oil.

Only three nations—the United States, Germany, and Japan—have sufficiently sturdy economic and financial systems to buy their way out. Their success would spell doom for their less powerful partners in the confederation of industrial countries. The lesser industrial powers, singly or in cooperation, would inevitably retaliate, using whatever means came to hand to avoid disaster. The less-developed countries, without the protection even of the present world economic order, could in desperation threaten the whole fabric of economic and political relations. The United States, if it indulged in an orgy of selfishness, could take the world back to the worst excesses of the age of mercantilism.

Soviet and Chinese troublemaking could make matters worse. Nothing would please the Soviet Union more than an atmosphere of conflict in the noncommunist world. After all, the communist doctrine has for decades been predicting the collapse of capitalism in fights over raw materials and markets. A punch here, a jab there, some arms for this or that country, the use of their own oil in another country, and the conflict heightens, opening the way for Soviet territorial and ideological expansion.

The world would be a mess, with civilized behavior a rarity. The present world order would not be able to stand up to it. Bad as it is, the present order follows certain rules of conduct that helps keep conflict within bounds. The market is one aspect of that order that, when functioning properly, helps to contain conflict. The breakdown of the oil market would dissolve some of the glue that holds the world together. No one knows where it would all end. Bare-knuckled conflict is a poor replacement for the market.

In their book, *The Pressures of Oil* (Harper & Row, 1978), Peter Odell and Luis Vallenilla eloquently plead for cooperation. They regard it as essential that the mechanisms for cooperation between

OECD, including the United States, and OPEC, on its own behalf and that of other less-developed countries, be established now to avert disaster. In the absence of such cooperation, "suspicisions and conflicts seem so likely to intensify as to give rise to serious doubts about the ability of the system to survive in a recognizable form."

WORLD COOPERATION

What of the possibility of cooperation? The cooperation of formal international conferences, with polite speeches, polite handshakes, and polite communiques is certainly not enough. Rather, cooperation would require the sweating, shirt-sleeved hammering out of detailed decisions that make a difference, and yet are decisions that nations can live with. It would demand a degree of cooperation that the world has not yet experienced. The closest thing to it has been wartime cooperation among allies. In a sense, it would be wartime cooperation. The enemy is the chronic shortage of oil that imperils all nations as well as the chaos of conflict.

The meeting of major oil-importing countries immediately after the 1979 mid-year price increase by OPEC may be the harbinger of cooperation among importers. The meeting, also midst a shortfall of crude-oil exports by OPEC and an acute gasoline shortage in the United States, agreed to set import quotas for each of the nations. The quota for the United States was 8.5 million barrels a day for the period up to 1985. This is much less than most of the projections of American demand. Other nations specified some growth of imports but at greatly reduced rates compared to historical trends. In the Tokyo meeting there was even some evidence of producer cooperation. The United States had informal and nonbinding assurances from Saudi Arabia that it would increase exports if the oil-importing countries took a hard line on oil conservation. How this incipient cooperation will work out remains to be seen, but it is at least a beginning.

Success would require cooperation at all levels and among and between every group of nations. No new institutions necessarily need be created, but effective use of existing ones, starting now, is mandatory. Both the less-developed and industrial countries must strengthen cooperation within their own groups. The Group of 77 and OECD must become more efficacious and direct more attention to oil and energy. Now is also the time to begin the first steps of cooperation among the three—OPEC, the Group of 77, and the OECD.

The first item on the agenda would be a deal between OPEC and the industrial world. The deal would necessarily include the less-developed countries as well. It would have two parts. One part would deal with oil. The other part must concern the reforms of the world economy. Real progress toward agreement, however, is improbable until the oil shortage becomes acute.

OPEC must then agree to supply a known amount of oil. It may be more than OPEC would prefer to supply, a strain on their capacity, and a more rapid expansion than members would prefer. The quid pro quo for the guaranteed supply would be a series of measures by the industrial countries that would assure stern conservation, a timetable for the introduction of substitutes, and a guarantee that the purchasing power of assets accumulated by members and their earnings would not deteriorate through inflation.

The exploration for new oil must be a cooperative matter. No country can regard its oil as a completely sovereign right, above the dictates of the world order. Many less-developed countries and some industrial countries have oil yet to be discovered. They need resources to explore for, find, and produce the oil. The advanced countries, through their oil companies, and perhaps even OPEC, can help do the job. It must be done in a way that not only produces the oil but also enhances the development prospects of nations where oil is found. Cooperative arrangements among buyers, sellers, and intermediaries of new oil must support the cooperation between importers and exporters.

The price and amount produced would be bargaining matters, bearing little relation to the market solution that cooperation replaced. The price must be high enough to cover the development needs of OPEC members and other exporters and yet low enough to permit buyers to pay without crippling their economies. The amount must cover only the most vital needs of importing countries, including the less-developed countries. The mutual guarantees would have to be strong enough that all participants would feel that no other arrangement would confer more benefits or require less sacrifices.

The agreement would be extremely difficult to negotiate, to implement, and then to keep on the track. Undoubtedly, conflicts would arise. The question is whether the world order is a cooperative one in which troublesome conflict occasionally erupts but is contained in the system, or is a conflictive one in which cooperation sometimes occurs. Burdensome and difficult as it is, cooperation would be better than the market and better than conflict.

The oil deal is only part of the package. OPEC and the less-developed countries will be in a position to insist that appended to agreements on oil would be the negotiation and implementation of reforms in the world economy. Many specific measures would be necessary. The developed and less-developed countries must come to terms on raw-material pricing, foreign aid commitments and capital availability, access to markets, technical assistance and technological transfer, debts, and other matters—the same items that were on the agenda at the Paris North-South Dialogue. This time it will not be enough to meet and talk. By agreements acceptable to all, the world must execute measures that redistribute the benefits of world production.

The future holds both cooperation and conflict over oil. The balance depends not on oil but on men and nations. Can the present OPEC cooperation and the cooperation among industrial countries be amplified into cooperation among all producers and consumers? If it can, cooperation may be able to serve the national interests of all countries better than unsatisfactory market arrangements or international economic conflict. It cannot serve these interests unless the interests of all, even the most powerful nations, include a recognition, acceptance, and allowance for the interests of all other nations. Oil may bring the world to the brink of disaster and the triumph of national self-serving or to a new day of international cooperation.

One key to the prospects for cooperation is the continued cooperation of the members of OPEC. Is OPEC stable or will it topple under the pressure of untoward events in the oil market and the hunt for economic gain? The next chapter addresses this question.

Chapter 11

OPEC Stability

The forces opposed to OPEC have predicted its collapse from the beginning. In the early 1960s the world neither noted nor took the organization seriously. When, along with rising demand, OPEC stirred up oil markets after 1970, companies and oil-importing countries still regarded it as a transitory but troublesome, challenge to their control of the market. Even after the price revolution of 1973, economists and politicians, looking at the unusual circumstances of OPEC's surge to power and the history of cartels, expected cooperation among the OPEC members to falter when the market returned to normal.

When OPEC did not stumble in the years right after 1973, OPEC watchers in the press and among politicians, economists, and energy specialists divided into two camps. The largest group continues to forecast its disintegration. Many members of this group seem motivated by wishful thinking and public posturing, rather than by a study of OPEC's strengths and weaknesses. The second group broke from the earlier consensus and now accepts OPEC as invincible and eternal, able to work its will when and as it chooses. This group, impressed by OPEC's success and the apparent impotence of the companies and consumers, is often interested in pounding home the lesson of OPEC—the scarcity and uncertainty of oil supplies.

Many forces work to fortify OPEC cooperation. Other forces eat at the vitals of its solidarity. OPEC could collapse tomorrow. It could also become a solid alliance of nations who greatly influence the world even after their oil disappears. OPEC's future depends on external events and circumstances as well as on internal tensions. The degree of its unity will wax and wane, and oil-importing countries will be able to influence its behavior. The glue that binds OPEC members together is stronger, however, than the solvents that erode their cohesion. OPEC will dominate the oil market for many years, barring cataclysmic events that alter all world economic and political relationships.

FEAR, PRIDE, AND PREJUDICE

Never forget that the nations of OPEC have been the underdogs for centuries. The Arabs, Persians, Africans, Asians, and Latin Americans have long lived under the white man's economic and political yoke and suffered the stripes of racism. They live in a world where the white man still controls most of the centers of power. Most members, apprenticed under white rule, have had little experience in nationhood. All were and are, despite the wealth dumped on them, poor and underdeveloped countries. They fear and distrust Europeans and Americans. A mixture of fear, pride, and prejudice solidifies their cooperation.

The taste of power accorded by OPEC's control of oil supplies gives members much to lose. They recognize that only special market conditions, the 1973 Arab-Israeli war, and their own cooperation placed them in the catbird seat. These conditions have now passed and only cooperation stands between these countries and economic and political disaster. No OPEC member can view calmly the reassertion of the companies' and consumers' dominance. That fear impels members to huddle together beneath their cloak of cooperation.

OPEC members are proud of their achievement. For the first time in history a group of "nothing" countries wrested power from the great nations. Not one of the members, not even Saudi Arabia, could have done it alone or even claim credit as the key country at the beginning of the 1973 takeover. Despite its preeminence, even today Saudi Arabia is not OPEC. The pride of accomplishment is mutually shared and will last only as long as OPEC members hang together. The loss of self-esteem, not to mention the economic and political penalties, would be a severe blow to the national pride of those OPEC members responsible for undoing their magnificent coup.

The force of these psychological forces, buried deep in the minds of the leadership and people of OPEC, is great. Their determination to share power and prestige with the white man fortifies the member's cooperation. OPEC members also like the attention of their new position, including the fawning of bankers, businessmen, and politicians of powerful countries. They take pleasure in their new ability to chastise the behavior of industrial countries, their wasteful opulence, and their dedication to goods and technology that make them dependent on oil. When world leaders call OPEC a cartel, misrepresent their actions, or employ policies that are hostile to their cooperation, the members bristle.

Members consider themselves beseiged by powerful forces. The companies and the oil-consuming countries—the greatest aggregation of wealth the world has ever known—are announced antagonists. Members believe that their legitimate actions are misunderstood and that their point of view does not receive a fair hearing in the press, diplomatic circles, and council chambers of the world. They think the people in oil-importing countries blame OPEC for the present oil situation. Beleaguered and defensive, OPEC members think that their foes eagerly await a split in their ranks and are ready to pounce on any misstep.

These intangible forces can also exercise a divisive influence. Racial and ethnic heterogenity, chauvinism, jealousy of the influence of other members, and fear of dominance by a single member take their toll. Saudi Arabian and Iranian leaders expressed amusement at Paul Erdman's book, *The Crash of '79*, a best-selling novel in which Iranian territorial designs lead to disaster for both countries. But as Anthony Sampson notes, the two largest producers are wary of one another and radical Arab countries—notably Algeria, Iraq, and Libya—are also leery of conservative Saudi Arabia (*Seven Sisters*, pp. 349-353).

Venezuela and Libya, once more powerful in OPEC councils, lament their diminished roles. A campaign issue in the 1978 Venezuelan presidential election was Venezuelan lack of aggressiveness and modest role in OPEC. Large countries, such as Saudi Arabia, become annoyed that small countries, like Gabon, exporting less than 3 percent that of Saudi Arabia, officially have the same voice. Members often criticize other members' policies and behavior as much as they do those of the industrial countries. Within OPEC, personalities often clash and offense is sometimes taken as ministers urge their own national positions.

Psychological costs must be weighed against benefits. The weight of these forces today favors continued cooperation. The enmities between oil-exporting and industrial countries are deep, long-standing, and have sharp edges. Since OPEC cooperation is new and was attended by a spectacular victory, members often tend to overlook the deficiencies of their allies. Members are still in euphoria, willing to ignore the internal psychological stresses. Over a longer period of time, depending on the balance of other forces, these kinds of internal dissensions could tip the scale in the other direction.

PRICE CONFLICTS

Since the unanimous decision of members sets the price of oil, it is a fruitful source of conflict. Some members in the late 1970s want large price increases. Others want moderate increases. Often a few, notably Saudi Arabia, want no price increase. Members bargain among themselves to determine the price. With one exception—the 1977 split between Saudi Arabia and United Arab Emirates and the rest of OPEC—members set a single price at the semiannual conferences. Still, controversies swirl around price and sometimes shake OPEC's serenity.

The controversies revolve around each member's view of the state of the market and the impact of a price change on its own economy as well as on the economies of industrial countries. The critical issue is the effect of price increases on members' revenues. A price increase of 10 percent might depress the market by only 7 percent, thus increasing members' revenues. But unless the demand by industrial countries grows, some OPEC members will have to reduce exports to sustain the price, and those members could lose revenues.

Who takes the cut? If Saudi Arabia, Kuwait, the United Arab Emirates, or others take the large cuts, they may earn less revenue, despite receiving more per barrel. In any case, the reshuffling of exports among members redistributes income within OPEC, benefitting some members more than others—or benefitting some at the expense of others. Without production controls to distribute revenues in an acceptable manner, price increases inevitably raise the market-share problem.

Because Saudi Arabia is the largest producer, with the greatest oil and financial reserves, other members often expect it to cushion the effect of increased prices. Saudi Arabia has used its swing position to do just that. Its willingness to do so vests it with a powerful voice in the price decision. To avoid absorbing the impact of price increases, Saudi Arabia has favored small, if any, price increases. Because of rising import prices and depreciation of the dollar, however, even Saudi Arabia's determination to keep prices down is wearing thin.

All OPEC members are greedy. They all want as much money as they can get, but members differ in how much money they need. Some—Libya, Saudi Arabia, and other Persian Gulf Arab members— have small populations, limited capacity to use money productively, and large reserves. For others—Iran, Iraq, Indonesia, Algeria, Nigeria—the reverse conditions prevail. Although Iran joined the price conservatives momentarily in 1977, the prototypes of the hawk and the dove in OPEC are still Iran and Saudi Arabia, respectively.

Misbehavior by members grasping for more money would come from the desperate need for money. The fourfold price jump in 1973-1974, however, endowed all members with more money than they ever dreamed of before. True, they became accustomed to it quickly and some have managed to spend all or nearly all of it. Members' financial needs reflect heightened ambitions and new plans, often exaggerated by the ecstacy of apparent plenty. Rather than threaten the integrity of OPEC, however, members will re-assess their needs, putting some off to another day or relying modestly on credit, which as long as they have oil, is good.

Some challenge this reasoning. Imports of OPEC have shot up faster than expected. The new plans call for galloping imports. Armaments, that bottomless pit into which money pours, has quietly consumed billions. Inflation in import prices, omissions from plans, waste, and undercosting of projects will raise OPEC financial needs so much, some argue, that not only will most members soak up current income, but they will also use up reserves and credit and still not have enough. The only solution is to sell more oil. Yet the need to sell oil will reduce OPEC's flexibility, according to this argument, and could threaten OPEC cohesiveness.

Uncertainty about future expenditures and the state of the oil market make it impossible to say who is right. The OPEC surplus declined and almost vanished as the oil market stagnated from 1974 through 1978. The tighter market beginning in 1979 is boosting revenues and restoring the surplus. The initial splurge of spending has instructed OPEC members that development is more than spending money. The rate of increase of imports, still high, is declining. It is becoming apparent to most members that the lack of labor and entrepreneurship, not the lack of money, bars development, and that development is a slow and arduous process. Members still have more than $160 billion in financial assets and no member is seriously in debt. The probabilities seem to lie with OPEC's retention of the ability to reduce production if demand falls, without crippling financial restraints.

Price increases could induce recession in industrial countries, reducing demand enough to threaten revenues. In effect, OPEC could miscalculate, damaging their customers so much that they damage themselves as well. Indeed, the Saudi Arabian fear of harmful effects on industrial countries in part motivates its conservative price policy. To Iran and some of the more alienated Arab countries, the concern over customers' plight is slight. But OPEC cannot ignore the recession of 1974-1975 and slow growth since. OPEC knows enough about the oil market, however, and the effect of its actions to try to avoid self-defeating measures.

Price conflicts and the bargaining method of resolving them do not threaten OPEC. If any unresolvable difference arises, OPEC either does not change the price at all or lets two prices prevail for a short time. Since slight differences exist in any case, the added divergence does not cripple the market, but it does pressure both sides to compromise. The next bargaining session unifies the price. The one instance of two-tier pricing did not alter the revenue shares radically but brought enough tension in the market to effect the compromise. With the preponderant influences within OPEC on the conservative side, no foreseeable price dispute seems capable of undermining the organization.

CHEATING TO INCREASE REVENUES

The hidden bomb in OPEC is the uncertain distribution among members of exports and revenues. In a cartel—which OPEC is not—the most meticulous attention goes to determining how much each enterprise shall produce, to the means for policing production, and to enforcing the quota agreement. Even so, economists argue that the demise of most cartels originates in the failure to observe their production quotas. Enterprises form cartels because acting together they can make greater profits and achieve other goals that no one firm can. But after the cartel is formed, firms find that by producing more than they agreed, they can make even more money.

In OPEC and in a cartel, the temptation to cheat is great. Historically, it may have overpowered some cartel enterprises. An enterprise never intends to break the cartel by exceeding its quota. It just wants to make a bit more money. The other enterprises who stay within their quota make a bit less. No matter how well the first enterprise guards its secret, another enterprise finds out. To defend itself, it also produces beyond its quota. Then another follows suit, adding more pressure. Finally, production has increased so much that the cartel price collapses and the cartel disintegrates. This standard analysis of cartel behavior, however, omits other perhaps more important reasons for the short life of cartels. Many cartels perform their function and outlive their usefulness, or members discover other means, usually informal, to accomplish their purposes.

Let there be no doubt; OPEC will collapse if its members do not control their own production. Should Libya, secretly or openly, expand its production significantly, eroding the revenues of others, the others will retaliate. The ensuing controversy could kill OPEC. Still other members, to maintain their own revenues, would increase exports also, shaving the price if necessary. The house of OPEC could not survive.

If they can earn more revenues, then why does not Iran or Venezuela increase exports? To OPEC members money isn't everything. It is important, but other goals and even a different interpretation of the revenue goal motivates them. Since 1970 OPEC members have made a lot of money. It came from cooperation within OPEC. If OPEC collapses, members fear that the money won't be there. OPEC solidarity comes first and money next.

The profits of cartels come from cooperation too; yet, they do not stick together. Why should OPEC not follow the cartel trail? The cartel enterprise wants money because moneymaking is often one of the most important reasons for its existence. But Iran, for example, wants not money, but the development of its economy, strong defenses, prestige, and the achievement of all manner of economic, political, and strategic aims. A nation and its people live forever and command a sense of history as well as loyalty from its leaders. But a business is a business.

OPEC can earn money from its oil now or later. The scarcity value of oil goes up as supplies diminish. OPEC believes its price can shadow this scarcity value. A cutback in production now may yield even more revenue later, perhaps more than revenue now would yield as an investment. Any glut in the market, as in 1975 and 1978, is temporary. Saudi Arabia and others whose capacity to absorb capital is limited sacrifice little or nothing and may even gain by keeping their oil in the ground.

The multiple goals of members water down their enthusiasm for money. What profit is it to Iran if it has a few billion more now, but at the turn of the century when its wells are dry, it still has an underdeveloped economy, has no friends and must bow to forces over which it has no control? Better to be a poor member of a powerful coalition than to be independently poor. Better as a member of OPEC to accomplish many goals, accepting financial compromises made necessary by cooperation, than to act like a greedy merchant.

The organization of the industry and its physical limitations dampen large-scale cheating. Despite its size and complexities, the companies and OPEC are well-informed about the most minute details of the market. Cheating could not go undetected a fortnight. The tell-tale computer trail of shipments, prices, deliveries, by port, company, grade, buyer and seller, would reveal all.

Capacity not used soon ceases to be capacity. Venezuela produced 3.7 million barrels a day—capacity production—in 1970. In gradually reducing production since, its capacity in 1978 was reduced to only 2.4 and its production to only 2.2 million barrels a day, although its oil reserves are up. Members must spend time and effort, change plans, and risk costly mistakes to maintain capacity and to expand even within capacity limits.

The absence of OPEC production controls may be an advantage. It is almost impossible for thirteen countries to agree on market shares. The elaborate formulae that take into account current and historical production, oil reserves, economic conditions, financial reserves, capacity, plans, and other factors inevitably work to the advantage of some members and the disadvantage of others. The temptation of the disadvantaged to cheat might be greater than the present voluntary system. It places greater responsibility on Saudi Arabia which, along with others such as Kuwait, can permit exports and revenues to vary without sacrifice. And it works, without the punishing ordeal of agreeing on and enforcing quotas.

OPEC has organized the crude-oil market up to this time so that members will supply whatever demand appears at oil's current price. Members retain the option to change the price, accepting the demand implications of the new price. If demand falls, OPEC must reduce supply. If demand rises, OPEC must increase supply. This residual-supplier stance of OPEC results from excess capacity. As members reach capacity limits in a few years, OPEC price and production strategy may change.

Although the potential exists for crippling OPEC from the demand side in the near future, those prospects are already slim and grow slimmer. The opportunity has probably passed. If in 1974 the United States and other oil importers had mounted an all-out effort to reduce oil consumption and had succeeded in drastically cutting demand, holding it down for a number of years, OPEC might not have been able to weather the storm. If the industrial countries had found non-OPEC oil or introduced substitutes quickly, along with the oil demand reduction, OPEC might have lost control permanently.

NON-OPEC OIL AND SUBSTITUTES

OPEC, of course, does not have a corner on world crude oil or on energy. Members produce only a little more than one-half of total world oil. The world's largest producer, the Soviet Union, produces about one-third of total OPEC production. The third largest producer, the United States, produces not quite as much as the second largest, Saudi Arabia. Both the Soviet Union and the United States, however, like many lesser producing countries, require all of their own production for domestic use.

OPEC dominates world crude-oil exports. In recent years members of OPEC have provided 80 to 90 percent of world imports. Non-OPEC oil production will not threaten the members' grip for many years, if ever. In order to make inroads, non-OPEC production would have to increase much faster than that of OPEC. New exporting countries would have to appear, or producing countries would have to replace imported OPEC oil with new production at home. The new exports, in substantial volume, would have to appear on the market at prices lower than those of OPEC. The new exports and the imports replaced by domestic production would also have to cut OPEC demand by enough to threaten the OPEC price, exacerbating the market-share problem.

Where is that oil going to come from? New fields could be discovered tomorrow to put OPEC out of business. But it isn't probable. The discovery of oil is not a haphazard affair. Certain physical conditions must prevail and the worldwide oil industry has a good idea where the oil might be. Nothing in the analysis of petroleum scientists and geologists indicates that with known or projected technology and costs, oil discoveries in the next few decades will upset the non-OPEC-OPEC balance.

The greatest hope of diminishing OPEC's influence through increasing non-OPEC production has centered in the North Sea, Alaskan and U.S. offshore fields, Mexico, China, and in the possibility of exploiting oil tar and sands. The North Sea will certainly make a difference for Great Britain who can supply its own needs into the early 1980s. Norway and other countries will also benefit. Production in the North Sea, however, will peak in the early 1980s, and by about 1985 Great Britain will be importing again. Norway's conservative production policy guarantees that little of its oil will see the market. Westerners estimate a great potential for China, but if it exists, China may decide to reserve its oil for its own industrialization and growth.

Alaskan oil and potential offshore U.S. production will disappear in the growth of U.S. demand. Not only has American production been declining since 1970, but its reserves have also declined, even in Alaska. All U.S. proven reserves stand at about nine years' supply. Alaska for a time will replace some OPEC imports, as production in the Atlantic may. But they will only slow the rate of increase of import demand. The American oil deficit will continue to help keep OPEC strong.

Oil experts have been excited about the new fields in Mexico. In just a few years, Mexico could become a major exporter. Could Mexico undermine OPEC? There are two problems. Much of the new Mexican oil has a very low recovery rate, so that while the amount may look impressive, the amount that can be extracted may well be quite modest. In addition, Mexico is becoming a large consumer. Mexico may decide to conserve its oil for its own needs. Such exports as take place may follow OPEC policies. U.S. opposition might make it uncomfortable for Mexico to join OPEC, but already its modest exports conform to OPEC price policy. Mexico may be a de facto OPEC member. Certainly, the world cannot count on Mexico opposing OPEC.

Oil outside of OPEC will not in the foreseeable future loosen OPEC's stranglehold. The OPEC share may decline, but the new oil will only lessen the degree of tightness in the market. Even if discoveries were made this year, it takes a decade or more to get the new oil to the market. When the oil is available, if it is, consumers will absorb it greedily with only a modest effect on the rate of increase in price.

The oil industry always reserves a caveat. Oil men have made discoveries where the experts confidently predicted oil did not exist. The proven reserves are estimates and probable reserves are guesses, and both are on the conservative side. Technology could make it possible to extract oil from places not thought possible today or to markedly improve the recovery rate. But the probabilities favor the continuation of the present geographic distribution of reserves and production. As reported in the *Christian Science Monitor* (September 29, 1978, p. 1), a recent Rand study confirmed the probable continued dominance of the Middle East in undiscovered oil.

Substitutes may eventually make OPEC irrelevant. One day OPEC oil and all other oil will dissipate into unrecoverable waste heat or will be husbanded for non-fuel uses. The work that is done by oil now will be done by substitutes. A vast array of substitutes await tapping. Other fossil fuels, such as coal, natural gas, shale oil, and tar can replace oil. Nuclear energy is in use on a limited scale, and can expand. The Edison effect may improve the efficiency of power production. The renewable resource, the sun, offers advantages. It is not a question of having enough energy. There is plenty. It is a question of having the right amounts of the right forms of energy at the right place at the right time.

Above all, it is a question of time and money. The OPEC oil price is well below the price of substitutes and will remain so for many years. The world will exhaust most of its oil first, then move on to the substitutes. Scientists and technologists speak of twenty years, forty years, or longer. The recent minibreakthrough in fusion energy establishes only one small part of its scientific feasibility. The price of oil will rise as it drains away, making the use of substitutes, regardless of costs and inconvenience, mandatory. In the meantime, OPEC remains strong and stable.

It is easy to forget that the problem the industrial countries are fighting is not OPEC, but rather their own increased oil demands and limited supplies. Even if OPEC had not come into existence and had not captured the oil market, the price would have risen, and oil-importing countries would be confronted with the necessity to conserve and produce more. All that OPEC did was to impose a particular institutional framework on the oil industry, remove decisionmaking from the industrial countries, and increase the price of oil more and sooner than would have been the case had the companies retained control.

In the late 1970s it is too late to fret over what might have been. The industrial countries, through slowing their demand, increasing non-OPEC supplies, and turning to substitutes, can only mitigate to some degree the effects of the growing scarcity. They no longer have the option of removing OPEC from the control of oil or of altering the structure of the oil industry. As each year passes without stern conservation measures by the United States and other oil importers, their position weakens and that of OPEC gains strength. The best that the industrial countries can hope for by their influence on oil supply and demand is to restrain OPEC's behavior.

THE OIL COMPANIES

The oil companies and members of OPEC were natural enemies in the past. For decades the exporting countries fought to enlarge their share of the oil dollar. The companies staved off the attack for many years to preserve profits and expand their markets, giving ground slowly and reluctantly. Despite the economic power of the companies, however, the oil-exporting countries have finally won. In the peace treaty, the exporting countries required that the companies cede nearly all their properties and control of price— moves that all but dealt the companies out of the crude-oil business.

Those who are looking for a countervailing power to OPEC often expect the companies to champion the cause of consumers and the industrial countries. The companies certainly have the power to disrupt the effectiveness of the OPEC market. The demand side of the crude-oil market is in their hands and the companies could create endless problems in the orderly marketing of OPEC's crude. They could demand participation in the price decisions and if OPEC did not give in, the companies could perhaps upset the market enough to make OPEC wish it had shared pricing with the companies.

OPEC acquired control of the market from the companies by the simple assertion of its power. The ploy worked because the companies, realizing that they no longer controlled production, accepted the price. Some people believe that if the companies and the consuming countries had refused to acknowledge OPEC's primacy, OPEC would have crumbled. Even today, the companies could, in concert or by following the industry's leaders, reassert the old posted pricing system or insist on sharing in the price decision. According to this view, much of the OPEC control is ephemeral and would evaporate if challenged.

The key to the control of price is production. At the same time OPEC took control of price, it was acquiring control of production. Some of this dominance came earlier through nationalization and regulation. The embargoing countries acquired complete control in the fall and winter of 1973 and other OPEC members, learning from their example, lost no time in acquiring the power to determine their production as well. As production passed into the hands of OPEC members, who would set the price was no longer in question. Although demand helps to determine price, so long as the exporting countries regulate supply to meet demand at their price, the demand side of the market is powerless to intervene without sacrifice.

Even if OPEC collapsed as a cooperative venture, the regulation of production would still remain the decision of the oil-exporting countries, not the companies. Outside the framework of OPEC these countries might behave differently. If each pushed production to capacity in the next few years and capacity production exceeded demand, the OPEC price could not stand. The companies would regain some of their command over the price. Some of the larger exporters—Saudi Arabia and Iran, for example—could, however, by restricting production, influence and perhaps determine the market price. Although nominally OPEC's disintegration would vest the demand side with more power, the companies could never regain their properties or the control of production. The fall of OPEC now does not imply the return of company power.

The companies help to keep the market share problem within bounds by preferential buying, in effect reinforcing OPEC stability. Before the takeover, individual companies produced, refined, and marketed the oil. The same companies continue to buy most of the crude that was once theirs. These preferential arrangements serve the mutual interests of the companies and OPEC. It means that if a company buys more or less oil from one member, the interests of that company in another member may be damaged. The preference system by which companies assure themselves of adequate crude supplies suffers. The companies have in fact an informal and flexible quota system for members, the breaching of which could hurt all companies and all members.

Unless the companies could fully restore their previous preeminence, they would prefer the present situation. OPEC lopped off a part of their business—crude oil—but left untouched their downstream activities. The companies continue to have access to crude oil, on a preferred basis, in the countries in which they operated at the time of the takeover. The market shares of the companies are undisturbed—an element of stability greatly desired by the companies. The OPEC price, although higher, adds to that stability. And the companies' actions support OPEC.

The higher price may impede the income growth of the companies' oil business. But the companies reason that even if they had remained in charge of the crude-oil industry, their price would also have advanced. The companies now endeavor to achieve their growth goals by moving into other energy fields and other businesses. The higher price has not disrupted the profitability of the companies. They make no money on the crude they lost, but they can make as much or more from refining, distributing, and marketing oil products, and in their other enterprises. Even though the payments of OPEC members for company assets are minimal, the companies continue to be as profitable, or even more so, than before.

The peace treaty between OPEC and the companies has become an alliance. OPEC has crude oil and is prepared to sell it. The companies sell oil products and needs crude oil to make them. The companies, of course, want to buy crude as cheaply as possible and OPEC wants to sell it for as high a price as possible. The companies recognize that OPEC now has the power to enforce its price. Shopping around and bargaining over price might save the companies pennies. It might also endanger the steady and reliable flow of crude oil, which is much more important to the companies. OPEC can assure the companies' stability and does not menace their profits or growth. Any course of action other than cooperation with OPEC would threaten their profits, stability, and growth.

THE CONSUMING COUNTRIES

In an all-out confrontation the oil-consuming countries could crush OPEC. Stern conservation measures, using oil only for its most vital uses, a large-scale exploration and investment program to find non-OPEC oil, and the rapid introduction of substitutes, could possibly so weaken OPEC that its price could not stand. The problem of the distribution of revenues among members would then rip OPEC apart. Such an endeavor would require faithful cooperation and active participation by all of the major oil-consuming countries, representing the highest priority of all their governments.

The cost would be staggering. The industrial countries would risk high inflation, inconvenience, regulation, and disruption for consumers and industry, as well as unemployment and loss of capital. The damage to the economies of the industrial countries might be greater than the damage to OPEC members and greater than the benefits they could achieve from lower oil prices for a period. The risks of disturbances, instability, and even war would be magnified. Many policies would be so repugnant and unnatural to most industrial countries that they could not be implemented. The industrial countries might not be willing to accept the risks and sacrifices for so limited a gain.

Besides their sharp influence on supply and demand of crude oil, the industrial countries possess other arrows in their quiver.

Although oil-consuming countries depend critically on OPEC oil, OPEC members also depend on the industrial countries to achieve many of their goals. OPEC needs a stable and peaceful world order and wants an international economic system that confers greater benefits on poor countries. The industrial countries only can guarantee peace and order and implement the changes they desire. The industrial countries can use their economic, political, and diplomatic posture and actions to influence, but not to destroy OPEC.

One measure, short of confrontation, proposes to manipulate the oil industry so that OPEC's control weakens. In the United States, for example, instead of relying on private companies to buy oil from OPEC members, a single government purchasing agent would solicit private bids from each nation. The orders would then go to the lowest bidders. In this way, each OPEC member is encouraged to enlarge its market share secretly by cutting the price. The OPEC price would tumble as shipments increased, and OPEC itself would crumble.

The trick would work only if demand were slack. The cohesion of OPEC is strong enough that the more probable outcome would be a series of identical bids. OPEC might even refuse to submit offers, risking the market for a time, but preserving their present market arrangements. In addition, the plan would supplant the oil companies as buyers of crude, controlling them to a degree that is onerous by present standards. The OPEC-breaking purpose of monopolistic buying is so obvious that only other strong pressures could induce OPEC members to fall into the trap.

The industrial countries provide the OPEC members with imports that their revenues buy. By refusing to export or by selectively controlling exports, the industrial countries might bend OPEC to their will, possibly even enough to make cooperation too costly to maintain. The United States, for example, could refuse to ship anything to an OPEC member unless it reduced the price of oil. The United States might refuse to ship arms, capital goods, or foodstuffs. Such obvious acts of economic warfare are a bit like driving a thumbtack with a sledgehammer and their clumsiness could have serious repercussions.

Political, economic, and diplomatic pressures by industrial countries could make OPEC members uncomfortable. Unless these pressures are in response to a specific OPEC action, however, the members can shrug them off as a part of the price of their dominance of oil. No specific action by OPEC members except another embargo is likely to offend the industrial countries so much that they would mount such an attack. If another embargo came, it would come not from OPEC as a whole, but from the Arab OPEC members, whose actions would be in response to a Middle Eastern war.

One ploy already in use by the United States and other oil-importing countries is to try to guaranty their oil supplies by currying favor with one or more OPEC members. American-Saudi Arabian relations improved steadily as a result of close economic relations, arms deals, and their mutual interest in Middle Eastern peace until the Egyptian-Israeli peace agreement. Saudi Arabian support of a comprehensive settlement only jeopardizes U.S. efforts to build bridges in the Middle East. U.S. support of the Shah of Iran proved a slender reed that now poisons its relationship with the new government. France and Great Britain, as well as other countries, have endeavored to strike up special relationships with OPEC members. Although these deals may be comforting to oil importers, they are uncertain. It is also improbable that oil-exporting countries could be induced to act against their own interests, such as the integrity of OPEC, to preserve the good opinion of some industrial country.

The industrial countries can amplify the effect of their own measures by acting in concert. One country's conservation counts for little. All undertaking heroic efforts together might reduce demand enough to trouble OPEC. One country's efforts to develop a substitute may be insufficient. All attacking the problem in a coordinated way could advance the timetable. But the means for international cooperation and coordination in energy policy at present are flimsy.

Only months after the price increase and embargo, the United States insisted on the formation of the International Energy Agency (IEA). Eighteen countries, with the notable exception of France, joined the United States in an effort to face another embargo by spreading the damage evenly, conserving enough oil to reduce demand and price, and discussing with producers an equitable price, market structure, and economic relations. The American government thought at first that it might be an effective counter-cartel of consumers. But IEA lacks supra-national authority, has no control over the policies of its members, and has spurred no action that its members would not have undertaken anyway.

Whether or not the beginnings of cooperation among oil importers apparent in the 1979 economic summit meeting are lasting enough to make a contribution depends less on international good will than it does upon the willingness of countries to take hard line domestic measures. Western Europe and Japan have shown some strength in conserving oil by substituting other energy sources and accepting very high oil-product prices. The United States continues to dally, and low energy prices still prevail. Draconian measures could reduce American demand for oil greatly. OPEC will modify its behavior not in accordance with pious resolutions at international meetings, but rather when oil-importing countries implement policies that affect their demand for oil.

The industrial countries may have the raw power to undo OPEC. But to do so would damage their economies and their image more than the gains that might attend OPEC's collapse. And no one knows what might succeed OPEC. Something even worse than OPEC for the oil-consuming countries could arise from the ashes of a successful assault.

Events since the early belligerent response to the price takeover and embargo indicate that the consuming countries, like the oil companies, have decided that cooperation with OPEC offers more long-run advantages and lower costs than confrontation. Slowing down the rate of increase of demand, increasing non-OPEC supplies, and searching for substitutes do not threaten the integrity or stability of OPEC. They do, however, restrain its behavior, as does gentle political and diplomatic pressure. OPEC suits the oil-consuming countries better than confrontation and its unpredictable results would. OPEC's knowledge that the industrial countries can introduce stronger measures and implement other debilitating policies encourages OPEC to refrain from taking full advantage of its position. Coexistence pays off for both sides, and the new recognition of interdependence stabilizes OPEC.

THE OPEC ORGANIZATION

Any cooperative venture needs an organization to define limits and responsibilities. The organization itself, however, may be a stumbling block. Members may object to one provision or another, to the degree of their influence, to benefits and costs, and to how the organization is run. Disaffected small members may quit, convinced that the organization does not pay them enough attention. Large members may quit in disgust at having to treat small members with kid gloves, believing that they can go it alone. What starts as high cooperation may bog down within the organization in bickering that endangers future cooperation.

OPEC has a primitive organization. Indeed, it could get along with almost no formal organization and secretariat. All the power of OPEC vests in the members, and not in the organization. The organization feeds information to members, does chores, conducts studies, provides a meeting place, services the conferences and other meetings, and tries to be useful. But it has no voice or power. OPEC organization is a shell.

Even the OPEC conference, where policy is set and decisions are made, has authority only insofar as it represents the view of member governments. All its actions are subject to ratification or rejection by governments. Unanimity is the rule, but no penalties exist for those who do not join the unanimous view. No member has ever withdrawn or been expelled. By failing to line up with all the others, a member signifies that it will not cooperate on a particular issue and that OPEC will not undertake it. Nominally, Saudi Arabia and Ecuador have the same authority and importance in the organization.

OPEC's simple organization serves the members better than a more sophisticated version. Weighted voting, vetoes by some, and elaborate rules do not guarantee cooperation, distribute benefits equitably, or assure participation. Indeed, they may have the opposite effect. In most organizations, a majority vote suffices to obligate those who voted against the measure. Vetoes by powerful members may stall measures that are in the interest of most participants. Secretariats often gain power and dictate the organization's policies. If such a system prevailed in OPEC, the chances for friction, discord, and breakdown would multiply.

Behind OPEC's formal organization, a shadow informal organization overcomes most of the demands of the unanimity rule. In fact, Saudi Arabia often possesses an informal veto because of its large production and ability to vary its production rate. Because other members expect Saudi Arabia to make most of the cuts if production must decline to sustain price, they defer to its judgment in price matters. Saudi Arabia is careful, however, not to offend other members who, although not critical in the world market, are yet part of the apparatus that conferred on Saudi Arabia such great benefits.

Iran has held Saudi Arabian influence in check. Before the revolution Iran pressed a more adventuresome role in OPEC. It was more needful of revenue, less concerned with the effect of price increases on industrial countries, and was not a party in the Arab-Israeli dispute. Its needs for arms in 1977 and 1978 and its shaky political situation tempered for a while that country's propensity for strong moves. The revolution, although it changed Iran's oil policy by reducing production and exports, did not disturb Iran's participation in OPEC. Since the revolution Iran has also resumed its hawkish ways on price.

Both Iran and Saudi Arabia have fickle fellow allies. Saudi Arabia can usually count on the support of the United Arab Emirates, and frequently it can count on Kuwait. Iran's voice is often amplified by Nigeria, Indonesia, and Venezuela, a bloc representing substantial crude production. The revolution and lower exports may reduce Iran's influence. Venezuela, with a new government in 1979, may assert itself more. Algeria, Iraq, and Libya, the radical Arab group, align themselves with those they think will benefit them by advocating higher prices.

Coalitions within OPEC are temporary and shift from time to time. On different issues, different groups of countries form blocs. Venezuela supports Iran sometimes and, then, Saudi Arabia on other occasions. Saudi Arabia and Venezuela are at odds on production controls, but that issue is now dying out. Small members join the winning side in a controversy in order to make the decision unanimous, even though a different decision might benefit them more. OPEC members change their positions to suit their own circumstances and their perceptions of the market and world conditions. If any trend exists, it is that most OPEC members increasingly share a common estimate of their own power and believe that the world must adjust to them. Saudi Arabia is becoming disgusted with the United States in oil and diplomatic matters, and less trustful of its leadership.

MEMBERS' CONFLICTING GOALS

OPEC is an instrument of each of its member's foreign and domestic policies. No conflict mars the pursuit of domestic policies by each member. OPEC serves to provide members with the funds to develop their economies, and no member begrudges the growth of other members' economies. No member wishes to see instability or problems in other members.

No member wants instability in the political regime of another member. Threats to the territorial integrity or domestic tranquility of a member could endanger cooperation. Saudi Arabia is acutely conscious that the internal problems visited upon Iran could visit Saudi Arabia next. Even when members use their funds for arms, no jealousy arises. The one possible exception is some future difference between Iran and Saudi Arabia concerning the Persian (or Arabian) Gulf. But for the foreseeable future, members are mutually self-supporting in their use of OPEC in domestic policies.

Even in foreign policies, members' attitudes in most cases are either indifferent or supportive. All OPEC members hope for world peace and stability. All members want a settlement in the Middle East. All want a loud voice in world councils. All want to see reforms in the way the world economy operates in order to benefit themselves as well as other less developed countries. None seek political or economic weakness in the industrial countries. None actively attempt to frustrate a member's efforts to achieve its foreign policy goals.

The general foreign-policy consensus conceals some conflicts and sharp differences in outlook. While none of the members are Communist countries, they run the gamut from sympathy to antipathy toward the Soviet Union. At one extreme stands Saudi Arabia, which regards the Soviet Union and Communism as the devil incarnate. Even though it has some misgivings about capitalist practices, Saudi Arabia, as represented by King Khalid and the ruling class, accepts capitalism as the best the world has to offer.

Other members have varying degrees of tolerance toward communism. Some trade with the Soviet Union and Eastern Europe, exchange diplomats, and even accept small domestic communist minorities. At the other extreme are three Arab countries—Algeria, Iraq, and Libya—that regard themselves as socialist, not Communist, countries. They often support Soviet policies and accept Soviet aid, including arms. But despite "playing footsie" with the Soviet Union, they are not satellites and remain strong Arab nationalists.

Minor Middle Eastern conflicts occasionally cloud the pacific OPEC horizon. Iraq has been the most troublesome country, often in conflict with Iran and Kuwait. Long standing territorial disputes erupt from time to time in an exchange of words, threats, and border troop movements. Then they blow over, without ruffling the feathers of OPEC. The Egyptian-Israeli peace agreement has created nominal unity among the other Arabs and a willingness to ignore many small differences, at least at the moment. But the anti-Egyptian Arab front is not solid. So far, however, internal Arab differences have not reached OPEC council tables. Except for Egypt, an underlying current of Arab cooperation is gaining ground, as witnessed by the recent formation of the Arab Monetary Fund.

Another explosion of the Arab-Israeli armed conflict could, however, blow OPEC apart. Seven of the thirteen members are Arab countries. Some of these eagerly seek the destruction of Israel. Others accept it grudgingly, do not wish it well, and have no fondness for Israel or its friends. All have an overarching goal: the rollback of Israel's borders to those that existed prior to the

1967 war. The non-Arab OPEC members do not always share the Arab states' antipathy toward Israel. For all but Iran, the dispute is far away and only an indirect threat. Many support the Arab states out of sympathy for their colleagues. Some OPEC members do sympathize with Israel and see an important role for it in the Middle East.

Iran's role in the Middle East is special. It is an Islamic country but is not Arabic. Before the revolution it sold oil to Israel and sometimes supported Israel in disputes with Arabs. Neither war, which might engulf Iran, nor complete peace, which might permit Arabs to turn their full attention to Iran, was pleasing to the shah. Since the revolution, Iran has adopted a pro-Arab stance. The ayatollah has eliminated oil exports to Israel and publicly supports the Palestine Liberation Organization. Still, Iran remains intensely nationalistic and shows no indication of full Arab support. The Arab countries, particularly Iraq and Saudi Arabia, take no joy in the revolution in Iran, fearing that the revolution could spread.

The separate peace between Egypt and Israel unsettles the other Arab countries, but has not disturbed OPEC. To avoid mounting tension in the Arab world, Saudi Arabia has lined up in opposition to Egypt and has reduced its aid to that country. The division in the Arab countries, however, does not threaten OPEC. All the Arab OPEC members oppose the separate peace. The danger to OPEC comes from the instability that the separate peace could introduce in the Middle East. If the Egyptian-Israeli peace is the first step in a general peace settlement, however, OPEC would eventually draw strength from it.

Renewed hostilities would put OPEC at hazard through another embargo. Although many industrial countries are now on much better terms with Arab countries than in 1973, a new embargo is possible. Many countries, including the United States, still support Israel. The United States continues to arm Israel and would step up arms deliveries if Israel faced military defeat. Although oil importers have more oil reserves now, and some depend less on Arab oil, the Arab countries, with greater command over the crude-oil industry, could now mount a more effective embargo. Once mounted, an embargo begins to lead a life of its own, the results of which are unpredictable.

The coming events in the oil market may have an effect something like that of an embargo. Nearly all estimates of increases in demand in the 1980s indicate that demand will exceed capacity production some time during the decade. The price must rise sharply to ration the limited supplies to oil importers. It is even possible

that OPEC members will decline to export oil beyond some defined limits, even if they have the capacity to expand further. They
may worry that demand-oriented production would lead to rapid
depletion of their oil reserves, saddle them with financial assets of
dubious value and earnings, and endanger their development
prospects. The oil market might be under such great pressure that
oil importers would seek a non-market solution.

The possible alternatives to the market are conflict and cooperation. Oil-importing countries could opt for economic, political,
diplomatic, or even military means to get the oil they need, in effect,
forcing OPEC to produce to meet their most urgent demands. Oil-
importing countries could also fall to fighting among themselves to
get enough oil, destroying the present cooperation and risking
world economic and political stability. OPEC cooperation might
survive external attacks and internecine struggles among importers.
In a different and possibly more chaotic world order, however, oil
and OPEC would still play vital, if altered, roles.

OPEC is vulnerable and at any time could fall. External pressures,
partly circumstantial and partly instigated deliberately by oil-
consuming countries, could undermine it. Internal dissension and
its own sins of omission and commission could also bring OPEC
down. It is unlikely that any single event or circumstance would be
sufficient. All of the elements of weakness are ever present and
OPEC lives with them. These weak elements could become more
important and could interact with one another enough to change
radically the world oil market again.

Nothing, however, compels OPEC's downfall. Its cohesion is
based on its success and the material benefits of cooperation, on
the internal strength of successful cooperation, on the luck of the
dice in the distribution of oil, and on the persistence of many of
the same external forces that favored OPEC's rise. Without fundamental changes in supply and demand conditions, in the world
economic and political situation, or internal squabbles, OPEC could
go on until its members run out of oil, or beyond. The probabilities
favor OPEC's survival with continuing benefits to its members for
the next two or more decades. The longer future contains many
imponderables and OPEC's favorable probabilities shorten. No one
certainly should hold his breath until OPEC collapses.

Chapter 12

OPEC's Place in the World

"Something new under the sun," His Excellency Ali Mohammed Jaidah, former secretary general of OPEC, told me, in explaining that OPEC, neither a cartel nor an oligopoly, was simply OPEC (Interview, *Worldview*, May 1979, p. 41). True, no organization quite like it has ever existed before. Although it has lame imitators, it stands alone as the first and only successful economic coalition of raw-material producing countries. Whether through high statecraft or tawdry blackmail, OPEC has undeniably shaken the world.

But has OPEC's existence and behavior in these years really changed anything? Is OPEC a flash-in-the-pan? Did it come, shining brightly for a moment, only to fade as the underlying forces of world economic and energy development reassert themselves? Will the United States and the great powers regain the upper hand? Will OPEC recede only to become a past moment of economic splendor for its members and a bad memory of transitory hard times for the industrial countries?

The answer is no to all these questions. OPEC has a permanent place in the world. For what it is—an economic organization setting the price and production of a vital commodity—OPEC has irrevocably bent historical trend lines and has changed forever energy and world economic relations. For what it represents—purposeful and thriving cooperation among a group of less-developed countries—OPEC is even more important. It presages a cooperative approach to world economic and energy problems.

Few comparisons exist. Perhaps the historic shift in United States' trade policy in 1934 as represented by the Reciprocal Trade Agreements, the formation in 1945 of the United Nations and its associated bodies, including the World Bank and the International Monetary Fund and the birth of the European Economic Community in 1958 set the precedents on which OPEC now builds. All of these cooperative ventures, including OPEC, have durably altered political and economic arrangements affecting every country in the world. All were new, different, and marked a turning point in world economic affairs.

By founding OPEC, setting its goals, and achieving those goals, the oil-exporting nations have dispelled for themselves and, indirectly, for all less-developed countries some common myths. Their behavior also has proved once again some old economic and political truths. Finally, the oil exporters have introduced some new and different elements into the international economic and political equation. As the time grows nigh when oil's scarcity impairs the operations of the oil market, OPEC will play one of the leading roles on the world stage, as victim, villain, or one of the good guys.

MYTHS EXPLODED

Gone is the myth that the white man must carry the burden for "lesser" breeds. The Japanese have disproved it in manufacturing production. Through OPEC, the Arabs, Africans, Latin Americans and Asians are disproving it in raw-materials production. OPEC members are operating one of the world's largest and most complex modern industries with efficiency and stability, yet tailored to the requirements of their cooperation. They are acquiring all the skills, from roustabout to managing director.

In technology and marketing, OPEC still depends upon the oil companies and consuming countries. In marketing, the use of the oil companies benefits OPEC and reduces duplication. Even so, the members are moving closer to the market and the consumer. Using present technology poses no problems for OPEC. In new technology not only OPEC but the entire world industry depends on a few large international companies. This dependence will endure.

Many cliches aver that the less-developed countries can meet and talk, but they cannot act together. OPEC has proved that disparate nations can cooperate effectively. A single common element can overcome language, religion, level of development, political and economic organization, history, and past conflicts among nations. Given the unifying element, less-developed countries can organize and innovate and can acquire and use power in their own interest as productively as any group of developed nations.

The OPEC organization is an ingenious method for resolving the conflict between cooperation and sovereignty. Unanimity preserves sovereignty. The loose and flexible organization, frequent exchanges and regular meetings, and detailed discussions of all matters associated with oil makes agreement possible, even though it may militate against the interests of some. OPEC cooperation did not stumble once it succeeded nor did power crumble in members' hands through national self-serving. OPEC has established that no group of nations has a monopoly on the ability to perceive and pursue self-interest through cooperation.

One great fear of the oil-consuming countries has been that if the less-developed countries acquired power and money, they would behave like small children. The stereotype would have these countries, inexperienced in handling wealth and influence, behaving selfishly and irresponsibly. But OPEC members, by the standards established by the developed countries, have neither abused their influence nor squandered their wealth.

Although members have served their own interests—that is, after all, the purpose of their cooperation—they have also shown sensitivity to the world economy and oil consumers. They have refrained from price increases that might damage others and have employed their money to support world money markets. OPEC's restrained behavior, impelled by self-interest, has established that its members, like developed countries, understand and act to preserve economic stability and order.

Newspaper tales of wild spending, sometimes true but grossly exaggerated, concern only a few people and small amounts of money. And their sumptious waste does not compare with the everyday profligacy of wealthy or even middle-class Europeans and Americans. Most of the OPEC money has paid for imports to support development programs. In questioning the wisdom of these programs, one really questions the ability and integrity of European and American advisors who recommended them. The arms expenditures of OPEC members do not begin to rival those of the great powers.

OPEC has invested its remaining funds, its surplus, in worldwide economic development. Most funds channel into world capital markets where they support the growth and prosperity of the developed countries. Its substantial foreign-aid programs, less than but rivalling those of the developed countries, support the economic development of the poor countries. Responsibility and concern for the international economic and financial system, whose soundness is transcendental to members' interests, keynote OPEC's new place in the world.

OPEC has exploded the myth that the less-developed countries are forever consigned a minor role in world affairs. No nation because it is poor, was a colony, or has little influence in world councils, need feel subservient to or in need of the tutelage of the great powers. One day, not far off, the rulers of Saudi Arabia, Iran, or other OPEC members will sit with the president of the United States and leaders of other developed countries to discuss world economic problems. Thirteen nations have cut the Gordian knot that has bound the use of power by small countries.

OLD TRUTHS

In addition to dispelling myths, OPEC has also reaffirmed some old truths that people often forget. For example, although the market system may be flawed, it is a remarkably tough and durable means for producing and distributing goods. Despite its intricacy, the capitalist system has withstood many great shocks, including prosperity and depression, inflation and deflation, growth and stagnation, unequal distribution of benefits, wars, and much meddling by governments and men dissatisfied with its working.

Those who feared that when OPEC increased the price of oil the world economy and finance would be at hazard worried needlessly. With all their deficiencies, markets, including the oil market, are the most efficient means for reflecting fundamental forces and for resolving economic conflicts. OPEC did not destroy the market. It used the market, just as the oil companies did before. The oil market now confers its benefits more to OPEC and less to the companies and consumers than it did, but it continues to function and will function for some years to come.

The world's fascination with widgets and gadgets—manufacturing production—often conceals the fundamental importance of raw materials. Without that which is on and in the earth, production does not take place. That's where everything starts, as OPEC has reminded the world once again. Technology, with all its wizardry, will never free mankind from dependence on the physical environment. Elementary observation, once the world's attention has been directed to resources, reveals that some raw materials which are the result of geologic evolution are limited in supply and are neither renewable nor reproducible.

The new twist in this old story is that in oil for the first time man has begun to approach the limits of a vital resource. For most minerals, land, and animals, supplies have been ample relative to the demands made upon them. In the singular case of oil, consumption outpaces discoveries and soon the consumption will begin its long glide down as substitutes, of necessity, replace oil. The only benefit to be derived from this occurrence, which will ineluctably raise costs, is that perhaps man will learn from it how to preserve better the remaining nonrenewable resources.

OPEC demonstrates the truism that power does not go forever unchecked. The existence of power, the power of the oil companies, for example, invited others to try to share that power and its benefits. Power begets countervailing power, as John Kenneth Galbraith reminds us in *American Capitalism* (Houghton Mifflin, 1956). When the oil companies and consumer interests dominated the crude-oil industry, the interests of owners provoked them to figure out ways to capture control, just as labor unions arose to participate in the profits of manufacturing industries. Company and consumer power begat OPEC countervailing power. That it has happened in oil should surprise no one, for it is one of the industries in which power has been most concentrated.

Nor should anyone be surprised that contention for ascendancy or, at least, a share of power in the world by less-developed countries will continue and broaden. Those denied even a mean living, as most of the world is, will seek and find ways to improve their lot. By linking oil to economic development, OPEC has provided the less-developed countries with a powerful level. With it, they expect to participate more fully in the economic benefits now so closely enjoyed by only a few countries and a small fraction of mankind.

Each developed nation has come to terms with the problem of unequal distribution of income within its own frontiers. By redistributing income and providing some economic benefits to all regardless of ability to pay, developed countries have supplied the model that the less-developed countries seek to establish on the international level. The sharing of benefits within the developed countries has moved forward slowly and peacefully, aided by orderly political procedures.

The international case is fraught with more hazards. The absence of world cooperation and the sense of world community inhibit the effort. Each nation divides people into "we" within its frontiers and "they" outside. And "they" are different and less important. Still, the less-developed countries propose to begin their campaign for more benefits now, aided by OPEC economic and political support and the example of its cooperation.

The world has entered the bargaining stage of the sharing of world economic benefits. The first effort—the Paris North-South Dialogue— ended in a stand-off because the developed countries perceived only high costs and few benefits and could not accept the methods proposed by less-developed countries. The lowered voices of these countries in the late 1970s does not imply a resolution of the conflict, but rather the gestation period during which the world adjusts to new power arrangements. In the 1980s when the oil crunch accelerates, other raw-material scarcities will arise, and as the concentrated power of today becomes more diffuse, the bargaining will begin again.

OPEC is responsible for the acceleration of the date at which the industrial countries must choose their course of action. Either they negotiate soon a higher level of economic benefits for the less-developed countries by some means and with perceptible progress, or they face an ugly confrontation in which everyone loses. The world of the poor cannot be put off, any more than the blacks and chicanos in the United States can be denied. If improvements do not come willingly, the less-developed countries will resort to their meager but growing countervailing power to get what they want.

OPEC has learned and teaches once again that cooperation pays off. What one country by itself cannot do, a group of countries with common interests and goals can do. OPEC is demonstrating also that the goals can be achieved without the loss of national identity. The whole can be greater than its parts and the parts need not dissolve into the whole. The little bits of power represented by each small nation can, when welded together, rival that of great nations.

Entertain no doubts, however, that even an army of small countries can ever really force the industrial countries to do what they are unwilling to do. The great powers with no difficulty could crush OPEC or any combination of small nations. The less-developed countries, like the OPEC members, must persuade the industrial countries that their greater participation confers benefits even on the powerful, that it does not threaten the vital interests of the developed countries, but that sharing benefits is a better solution than the consequences of either the status quo or confrontation.

OPEC further undermines the tired idea of determinism in world affairs. Great economic forces do not foreordain all events. What if Exxon had not cut its price in August 1960? What if those angry men, Dr. Perez Alfonzo and Sheikh Tariki, had not formed a friendship and dreamed a dream? What if the Shah of Iran, not always too fond of Arabs, had not joined? What if Colonel Qadaffi had not wanted to buttress his revolution with quick victories over foreigners? What if Sheikh Yamani had not staked out the posted price as central to OPEC policy in 1968? What if the harried negotiators for the oil companies had risked a world crude shutdown at Teheran in 1971? The past did not ride the waves of overpowering trends. It was made to happen by men and institutions. In the oil industry OPEC made its own history.

THE NEW AND DIFFERENT

Since the advent of OPEC, some things are new and different. All of the less-developed countries have changed the way in which they think about themselves and their relations with developed countries. The wave of nationalism has by no means receded, but cooperation among nations now has a higher priority. What one group such as OPEC can achieve through cooperation, others might also achieve if not in quantity, then at least in kind.

No longer is it one poor nation pleading for foreign aid as a favor from its former mentor or a superpower. It is now a bloc of countries demanding their economic rights from the countries that made the rules the world economy plays by. That the less-developed countries have not yet succeeded in achieving their goals does not belittle the importance of the change in their thinking. It is already forging new attitudes, policies, and actions, both in less-developed and developed countries. The four-fifths of humanity no longer will accept without protest the present distribution of economic gain that favors so much the dominant one-fifth.

It has not escaped the attention of either the powerful or the powerless that thirteen small and poor countries through cooperation and the exercise of national authority have imposed their will on the great powers. Such an event is not supposed to happen in the present world order. It had never happened before. Now that it has, the world is witnessing the unusual circumstances of the great powers adjusting their behavior and policies to confirm to a reality not of their own making. No longer can any American or European regard his country's actions as independent of the larger world community.

Before 1973 the United States and other developed countries spoke not of their dependence on other countries, but rather more gently of interdependence. Industrial countries, in fact, through their large enterprises and dominance of world markets, were almost independent of raw-material supplying countries. Since 1973, however, the industrial countries have voiced loudly their dependence on foreign supplies, the same complaint uttered by less-developed countries for decades. OPEC has forged genuine world interdependence, that is, mutual dependence that none can escape.

OPEC has redistributed world power. Before OPEC, the experts spoke of superpowers and two centers of political and economic strength. The Third World was a poor third, almost without influence. The recovery of Europe and Japan has diffused the American power center, but the United States remains the focus of economic, political, and military might opposed to the Communist world. The ascent of OPEC has now endowed the Third World with greater power and has created a Fourth World, the OPEC world, still aligned with the group from which it sprang.

The center of OPEC power, the Persian Gulf and Mediterranean countries, has become a formidable bloc, to be consulted in nearly all world matters. Saudi Arabia and Iran have pole-vaulted from lowly status to intermediate world powers. Nigeria in Africa and Venezuela in Latin America have become leading countries in their own regions. The simple two-center world is gone, replaced by a more complex arrangement of nations and blocs. The new distribution of power promises more stability than the old one with greater expectation of a more equitable sharing of economic benefits.

The members of OPEC have put to shame the industrial countries in foreign aid. The assistance of the United States, Western Europe, and Japan appears niggardly compared to the efforts of OPEC. The idea of less-developed countries helping other less-developed countries is novel enough. That its magnitude is many multiples of that of industrial countries on a relative income basis undermines the assumed right of industrial countries to regard their aid as entitling them to special privileges or to be regarded as the voice of authority in development matters.

OPEC aid, although more modest than that of industrial countries, is unfettered by national political self-interest and is unattached to economic benefits for the donor countries. Self-interest still underlies the aid, since through it OPEC as a group hopes to maintain the support of the Third World. OPEC aid also employs new devices to make its aid prompt and effective, including cooperating with existing world and regional organizations. Although OPEC aid will not resolve the development problem, its existence and magnitude demonstrates the spirit of cooperation needed if the less-developed countries are to progress.

It is fair to say that OPEC brought the world's attention to the energy problem. Specialists, of course, have known for many years that serious adjustments eventually awaited the heavy energy users because of limited oil supplies. But only the OPEC revaluation of oil in 1973 made the problem immediate, universal, and demanding of attention now.

As a result of OPEC actions every nation in the world is now re-examining its use of oil and energy. All countries are modifying their consumption and production patterns, altering their energy policies, and seeking to introduce substitutes for oil. By insisting on what they consider an appropriate payment for their oil, OPEC has lengthened oil's life and contributed to a less turbulent transition to other energy forms.

OPEC has made the case more eloquently than all the economics textbooks that the market is no respecter of persons or nations. Economists know this and those who favor the capitalist system can only hope that most markets will operate equitably most of the time for all concerned. They also hope that economic growth—made possible and stimulated by markets—will improve the lot of all.

Still, the basic flaw—that the market serves those who control it—persists. Most payments in the market system measure correctly the worth of the product and the productive contribution of those responsible for producing it. "The laborer is worthy of his hire," says the Bible, and the owner of capital, the *Wall Street Journal* echoes, must receive payment consistent with the contribution of his capital.

One class of payment escapes what most would consider a fair and well-defined return. In the case of unusual talent, scarce resources, or concentrated power, the owner or supplier is often able to receive a payment far beyond the legitimate contribution of what he owns or controls. Americans seldom cavil at the extraordinarily high payments to Robert Redford, Pete Rose, Pele, and Vladimir Horowitz. Many whose talent makes them unique resources and endows them with economic power have the ability to extract payments that are indeterminate in the market. Just so, those who own or control a unique and limited resource, such as oil, can receive as much or as little compensation as their power permits.

With the oil companies and consumers in control of the market, the owners complained that their payment was unfair. Consumers of oil products accepted the price as eminently just. With OPEC in control, consumers complain that the price is unfairly high, but the owners understandably consider their payment as appropriate. The market cannot resolve the issue of equity. Like the computer, the market grinds out the solution, responding to the forces put into it, without regard to equity.

PROGNOSIS

The present oil market serves its new masters—the members of OPEC—just as the oil market served its old masters—the oil companies and consumers. The rising demand and limited supply of oil, however, may create an oil market that in a score of years will serve neither. With chronic short supply, the market may need help in determining production, allocating it, and setting the price. The market, working alone, may produce results that damage buyers through unacceptable financial strains and sellers by exhausting their oil in exchange for unwanted financial assets.

Either conflict or cooperation will supplement the market. Some of the industrial countries may be tempted to wrest by force the oil from OPEC members. OPEC members would not accept such a move without fighting. Alternatively, the industrial countries may fall to squabbling among themselves over the oil available. The conflicts over oil in the event of a market breakdown would threaten the world economic and political system.

Cooperation among the oil-importing countries and between them and the members of OPEC offers more promising prospects. Cooperation is a more difficult route, calling for a higher degree of leadership and statesmanship. Each party to oil transactions would be required to accept sacrifices and forego benefits in the interest of achieving agreements acceptable to all. Although cooperation may allow continued economic progress for all countries, the future admits no escape from higher energy costs and unpleasant adjustments as one of earth's major fuels lapses into desuetude.

OPEC—bigger than oil and more important than thirteen countries—is an experiment in cooperation that invites the possibility of a negotiated solution to the problems of energy and economic development.

Bibliography

Abolfathi, F.; G. Keynon; M.D. Hayes; L.A. Hazelwood; and R. Crain, *The OPEC Market to 1985*. Lexington, Mass.: Lexington Books, D.C. Heath, 1977.

Acosta Hermosa, Eduardo. *Analisis Histórico de la OPEP*. Mérida: Universidad de Los Andes, 1969.

Acosta Hermoso, Eduardo. *La Comisión Económica de la OPEP*. Caracas: Editorial Arte, 1971.

Adelman, M.A. "Is the Oil Shortage Real." *Foreign Policy*, No. 9, Winter 1972-1973, pp. 69-107.

Adelman, M.A. "The Multinational Corporation in World Petroleum," in *The International Corporation*. Edited by Charles P. Kindelberger. Cambridge, Mass.: MIT Press, 1974.

Adelman, M.A. "Politics, Economics, and World Oil." *American Economic Review*, May 1974, pp. 58-67.

Adelman, M.A. "The World Oil Outlook." *Natural Resources and Economic Development*. Edited by Marion Clawson. Baltimore: Johns Hopkins University Press, 1964.

Adelman, M.A. *The World Petroleum Market*. Baltimore: Johns Hopkins Univeristy Press, 1972.

Akins, James. "This Time the Wolf Is Here." *Foreign Affairs*, Vol. 51, No. 3, April 1973, pp. 472-490.

Allen, Loring. "Not So Wild A Dream: OPEC's Amazing Rise to World Economic Power." *Harvard Magazine*, Vol. 80, No. 5, May-June 1978, pp. 22-28.

Allen, Loring. "Oil and Economic Reform." *OPEC Review*, Vol II, No. 5, December 1978, pp. 68-79.

Allen, Loring. "OPEC Speaks Out: An Interview with Ali M. Jaidah." *Worldview*, Vol. 22, No. 3, March 1979, pp. 41-46.

Allen, Loring. *Venezuelan Economic Development: A Politico-Economic Analysis*. New York: JAI Press, 1977.

Allvine, Fred C. *Highway Robbery: An Analysis of the Gasoline Crisis*. Bloomington; University of Indiana Press, 1974.

Al-Otaiba, M.S. (Saeed Al-Otaiba, Mana). *OPEC and the Petroleum Industry*. New York: Halsted Press, 1975.

Amuzegar, Jahangir. "The Oil Story; Facts, Fiction, and Fair Play." *Foreign Affairs*, Vol. 51, No. 4, July 1973, pp. 676-689.

Anthony, John Duke, ed. *The Middle East: Oil, Politics, and Development*. Washington: American Enterprise Institute for Public Policy Research, 1975.

Armstead, H. Christopher, ed. *Geothermal Energy*. New York: UNESCO, 1973.

Barrows, Gordon H. *The International Petroleum Industry*. New York: International Petroleum Institute, 1965.

Bergsten, C. Fred, ed. *The Future of the International Economic Order: An Agenda for Research*. Lexington, Mass.: Lexington Books, D.C. Heath, 1973.

Bergsten, C. Fred. "New Era in World Commodity Markets." *Challenge*, September/October 1974, pp. 34-42.

Bergsten, C. Fred. "The Threat from the Third World." *Foreign Policy*, No. 11, Summer 1973, pp. 102-124.

Bergsten, C. Fred. "The Threat Is Real." *Foreign Policy*, No. 14, Spring 1974, pp. 84-90.

Bergsten, C. Fred. "The Response to the Third World." *Foreign Policy*, No. 17, Winter 1974-1975, pp. 3-34.

Bill, James A., and Carl Lerden. *The Middle East: Politics and Power*. Boston: Allyn and Bacon, 1974.

Blair, John M. *The Control of Oil*. New York: Pantheon Books, 1976.

Blitzer, Charles; Alex Meerhaus; and Andy Stontjesdijk. "A Dynamic Model of OPEC Trade and Production." *Journal of Development Economics*, Vol. 2, No. 4, December 1975, pp. 319-335.

Bobrow, Davis B., and Robert T. Kudrle. "Theory, Policy, and Resource Cartels: The Case of OPEC." *Journal of Conflict Resolution*, Vol. 20, No. 1, March 1976, pp. 3-56.

Bohi, D.R., and M. Russell. *Limiting Oil Imports*. Baltimore: Johns Hopkins University Press, 1978.

Bohi, D.R., and M. Russell. *U.S. Energy Policy*. Baltimore: Johns Hopkins University Press, 1975.

Campbell, Robert W. *The Economics of Soviet Oil and Gas*. Baltimore: John Hopkins University Press, 1968.

Choucri, Nazli. *International Politics of Energy Interdependence*. Lexington, Mass.: Lexington Books, D.C. Heath, 1976.

Committee for Economic Development. *Achieving Energy Independence*. New York: Committee for Economic Development, 1974.

Connelly, Philip, and Robert Pealman. *The Politics of Scarcity*. London: Oxford, 1975.

Cremer, Jacques, and Martin Weitzman. "OPEC and the Monopoly Price of Oil." *European Economic Review*, Vol. 8, 1976, pp. 155-164.

Darmstader, Joel. *Energy in the World Economy*. Baltimore: John Hopkins University Press, 1971.

Davis, David H. *Energy Politics*. 2d ed. rev. New York: St. Martin's Press, 1978.

Doran, Charles F. *Myth, Oil and Politics*. New York: Free Press, 1977.

Doran, Charles F. *Dialogue on World Oil*, Conference on World Oil Problems, National Energy Project. Washington, D.C.: American Enterprise Institute for Public Policy Research, 1944.

Ebel, Robert E. *Communist Trade in Oil and Gas*. New York: Praeger, 1970.

El Mallaki, Ragaei, and Carl McGuire. *Energy and Economic Development*. Boulder: International Research Center for Energy and Economic Development, 1974.

Eckbo, Paul Leo. *The Future of World Oil*. Cambridge, Mass.: Ballinger, 1976.

Enders, Thomas O. "OPEC and the Industrial Countries, The Next Ten Years." *Foreign Affairs*, Vol. 53, No. 4, July 1975, pp. 625-637.

Energy Policy Committee of the Ford Foundation. *A Time to Choose: America's Energy Future*. Cambridge, Mass.: Ballinger, 1974.

Engler, Robert. *The Brotherhood of Oil*. Chicago: University of Chicago Press, 1977.

Engler, Robert. *The Energy Question, An International Failure of Policy*. Toronto: University of Toronto Press, 1974.

Ezzati, Ali. *World Energy Markets and OPEC Stability*. Lexington, Mass.: Lexington Books, D.C. Heath, 1978.

Fallman, S. David. *Energy: The New Era*. New York: Vintage Books, 1974.

Farmanfarmaian, Khodadad and others, "How Can the World Afford OPEC Oil." *Foreign Affairs*, Vol. 53, No. 2, January 1975, pp. 201-222.

Fischer, D.; D. Gately; and J.F. Kyle. "The Prospects for OPEC: A Critical Survey of Models of the World Oil Market." *Journal of Development Economics*, Vol. 2, December 1975, pp. 363-386.

Frank, Helmut J. *Crude Oil Prices in the Middle East: A Study in Oligopolistic Price Behavior*. New York: Praeger, 1966.

Frankel, Paul H. *The Essentials of Petroleum: A Key to Oil Economics*. London: Chapman and Hall, 1946, 1969.

Frankel, Paul H. *Mattei: Oil and Power Politics*. New York: Praeger, 1966.

Frankel, Paul H. *Oil: The Facts of Life*. London: Weidenfeld and Nichelson, 1962.

Fried, Edward R. and Charles L. Schultze, eds. *High Oil Prices and the World Economy: The Adjustment Problem*. Washington, D.C.: Brookings Institution, 1975.

Gardner, Richard N. "The Hard Road to World Order." *Foreign Affairs*, Vol. 52, No. 3, April 1974, pp. 556-576.

Gebelein, C.A. "The Effect of Conservation on Oil Prices." *Journal of Energy and Development*, Vol. 1, No. 1, 1975, pp. 70-93.

Ghadar, Fariborz. *Evolution of OPEC Strategy*. Lexington, Mass.: Lexington

Books, D.C. Heath, 1977.

Halacy, D.S., Jr. *Earth, Water, Wind, and Sun.* New York: Harper & Row, 1977.

Hart, Susan. *Producer Controls: Threat or Opportunities.* London: Commonwealth Industries Association, 1975.

Hartshorn, Jack Ernest. *Oil Companies and Governments.* London: Faber and Faber, 1966.

Hartshorn, Jack Ernest. *Politics and World Oil Economies.* New York: Praeger, 1962.

Hartshorn, Jack Ernest. *Objectives of Petroleum Exporting Countries.* Nicosia; Middle East Petroleum and Economic Publishers, 1978.

Heller, Charles A. "Ten Years of OPEC." *World Petroleum,* Vol. 41, 1970, pp. 46-54.

Hirst, David. *Oil and Public Opinion in the Middle East.* London: Faber and Faber, 1966.

Houthakker, Hendrik S. *The World Price of Oil: A Medium Term Analysis.* Washington, D.C.: American Enterprise Institute for Public Policy Research, 1976.

Ignatius, Miles. "Seizing Arab Oil: The Case for U.S. Intervention." *Harpers,* March 1975, pp. 45-62.

Inglis, K.A.D., ed. *Energy: From Surplus to Scarcity?.* New York: Wiley, 1974.

Iskander, Marwan. *The Arab Oil Question.* 2d ed., rev. Benoit, 1973.

Issawi, Charles, and Muhammed Yeganeh. *The Economics of Middle East Oil.* New York: Praeger, 1962.

Itayim, Faud. "The Organization of Petroleum Exploiting Countries." *Middle East Forum,* Vol. 38, December 1962, pp. 13-19.

Jacoby, Neil H. *Multinational Oil.* New York: Macmillan, 1973.

Kess, Malcolm H. *The Arab Cold War 1958-1967.* 2d ed., rev. London: Oxford University Press, 1967.

Kalymon, B.A. "Economic Incentives in OPEC Oil Pricing Policy." *Journal of Development Economics,* Vol. 2, December 1975, pp. 337-362.

Kennedy, Michael. "An Economic Model of the World Oil Market." *The Bell Journal of Economics and Management Service,* Vol. 5, No. 2, 1974, pp. 540-577.

Klehanoff, Shoohana. *Middle East Oil and U.S. Foreign Policy: with Special Reference to the U.S. Energy Crisis.* New York: Praeger, 1974.

Knowles, Ruth Sheldon. *America's Oil Famine.* New York: Coward, McCann, and Geoghegan, 1975.

Kraar, Louis. "OPEC Is Starting to Feel the Pressure." Fortune, May 1975.

Kraft, Joseph. "Letter from OPEC." *New Yorker,* 20 January 1975, pp. 64ff.

Krueger, Robert B. *The United States and International Oil* (report prepared for Federal Energy Administration). New York: Praeger, 1975.

Kubbah, Abdul A.D. *OPEC Past and Present.* Vienna: Petro-Economic Research Center, 1974.

Leeman, Wayne A. *The Price of Middle East Oil.* Ithaca: Cornell University Press, 1962.

Lenczewski, George. *Oil and State in the Middle East.* Ithaca: Cornell University Press, 1960.

Levy, Walter. "Oil Power." *Foreign Affairs*, Vol. 49, July 1971, pp. 652-668.

Levy, Walter. "World Oil Cooperation or International Chaos." *Foreign Affairs*, Vol. 52, No. 4, July 1974, pp. 690-713.

Levy, Walter. *Future OPEC Accumulating of Oil Money: A New Look at a Critical Problem*, Walter Levy Consultants, June 1975.

Limaze, D.R. *Energy Policy Evaluation*. Lexington, Mass.: Lexington Books, D.C.Heath, 1974.

Longrigg, Stephen Hemsley. *Oil in the Middle East*. New York: Oxford University Press, 1954.

Lufti, Ashnof T. *OPEC and Its Problems*. Beirut: Beirut Institute of Economic and Social Planning in the Middle East, October 1967.

Lufti, Ashnof T. *OPEC Oil*. Beirut: Middle East Research and Publishing Center, 1968.

Magnus, R. "Middle East Oil and the OPEC Nations." *Current History*, January 1976, pp. 22-26.

Mancke, Richard B. *The Failure of U.S. Energy Policy*. New York: Columbia University Press, 1974.

McCracken, Paul W. moderator. *Is the Energy Crisis Contrived?*. Washington, D.C.: American Enterprise Institute for Public Policy Research, 1974.

Maddox, John. *Beyond the Energy Crisis*. New York: McGraw-Hill, 1975.

McKie, James W. "The Political Economy to World Petroleum." *American Economic Review*, Vol. 64, No. 2, May 1974, pp. 51-57.

Medvin, Norman. *The Energy Cartel: Who Runs the American Oil Industry?*. New York: Vintage, 1974.

Menderhausen, Horst. *Coping with the Oil Crisis: French and German Experiences*. Baltimore and London: John Hopkins University Press for Resources for the Future, 1976.

Mikdashi, Zuhayr. *The Community of Oil Exporting Countries*. Ithaca: Cornell University Press, 1972.

Mikdashi, Zuhayr. *A Financial Analysis of Middle Eastern Oil Concessions 1960-1965*. New York: Praeger, 1966.

Mikdashi, Z.; S. Clelund; and Ian Seymour, eds. *Continuity and Change in the World Oil Industry*. Beirut: Middle East Research and Publishing Center, 1970.

Mikesell, Raymond F., ed. *Foreign Investment in Petroleum and Mineral Industries*. Baltimore: John Hopkins University Press, 1971.

Miller, Roger L. *The Economics of Energy*. New York: Morrow, 1974.

MIT Energy Laboratory Policy Study Group. *Energy Self-sufficiency: An Economic Evaluation*. Washington, D.C.: America Enterprise Institute for Public Policy Research, 1974.

Mitchell, Edward J. *Dialogue on World Oil*. Washington, D.C.: America Enterprise Institute for Public Policy Research, 1974.

Monroe, Elizabeth, and Robert Mabro. *Oil Producers and Consumers: Conflict or Cooperation*. New York: American Universities Field Staff, 1974.

Mosley, Leonard. *Power Play: Oil in the Middle East*. New York: Random House, 1973.

Nakhleh, Emile A. *Arab-American Relations in the Persian Gulf*. Washington,

D.C.: American Enterprise Institute for Public Policy Research, 1973.

O'Connor, Harvey. *The Empire of Oil*. New York: Monthly Review Press, 1955.

O'Connor, Harvey. *World Crisis in Oil*. New York: Monthly Review Press, 1962.

Odell, Peter R. *Oil and World Power: Background to the Oil Crisis*. 3d ed., rev. New York: Taplinger, 1975.

Odell, Peter, and Luis Vallenilla. *The Pressures of Oil*. New York: Harper & Row, 1978.

Organization for Economic Cooperational Development. *Oil: The Present Situation and Future Prespects*. Paris: OECD, 1973.

Organization of Petroleum Exporting Countries. *Sources of Petroleum Statistical Information*. Vienna: OPEC, 1966.

OPEC. *Annual Review and Record*.

OPEC, *Annual Statistical Bulletin*.

OPEC, *OPEC Bulletin* (weekly).

OPEC, *OPEC Review* (quarterly).

Pearson, S.R. *Petroleum and the Nigerian Economy*. Palo Alto: Stanford University Press, 1970.

Penrose, Edith. *The Growth of Firms, Middle East Oil, and Other Essays*. London: Cass, 1971.

Penrose, Edith. *The Large International Firm in Developing Countries: The International Petroleum Industry*. London: Allen and Unwin, 1964.

Penrose, Edith. "Profit Sharing between Producing Countries and Oil Companies in the Middle East." *Economic Journal*, Vol. 69, June 1959, pp. 238-254.

Pérez, Alfonzo, Juan Pablo. *El Pentágono Petrolero*. Caracas: Editorial Revista Politica, 1967.

Pollack, Gerald A. "The Economic Consequences of the Energy Crisis." *Foreign Affairs*, Vol. 52, No. 3, April 1974, pp. 452-471.

Rocks, Lawrence, and Richard P. Runyon. *The Energy Crisis*. New York: Crown, 1972.

Rifai, Taki. *The Pricing of Crude Oil: Economic and Strategic Guidelines for an International Energy Policy*. New York: Praeger, 1974.

Rifai, Taki. *International Petroleum Encyclopedia, 1973*. Tulsa: The Petroleum Publishing Company, 1973.

Rouhani, Fuad. *A History of OPEC*. New York: Praeger, 1971.

Rustow, Dankwart A., and John F. Mugno. *OPEC: Success and Prospects*. New York: New York University Press for the Council on Foreign Relations, 1976.

Safer, Arnold. *International Oil Policy*. Lexington, Mass.: Lexington Books, D.C. Heath, 1979.

Sampson, Anthony. *The Seven Sisters: The Great Oil Companies and the World They Shaped*. New York: Viking Press, 1975.

Sayed, Mustafa. *L'Organisation des Pays Exportateurs de Petrole*. Paris: Infirmerie Nationale, 1967.

Schutt, Sam, and Paul Homan. *Middle Eastern Oil and the Western World*. New York: American-Elsevier, 1971.

Shaffer, Edward H. *The Oil Import Program of the United States: An Evaluation*. New York: Praeger, 1968.

Shwadran, Benjamin. *The Middle East, Oil, and the Great Powers.* 3d ed., rev. New York: Wiley, 1973.

Solberg, Carl. *Oil Power: The Rise and Imminant Fall of An American Empire.* New York: Mason/Chaster and New American Library, 1976.

Stocking, George. *Middle East Oil.* Nashville: Vanderbilt University Press, 1970.

Stobaugh, Robert, and Daniel Yergin. *Energy Future.* New York: Random House, 1979.

Stone, Christopher D., and Jack McNamara. "How to Take on OPEC." *New York Times Magazine,* 12 December 1976, pp. 20, 47-55.

Szyliowica, Joseph S., and Bard O'Neill. *The Energy Crisis and U.S. Foreign Policy.* New York: Praeger, 1975.

Tanzer, Michael. *The Political Economy of International Oil and the Underdeveloped Countries.* Boston: Beacon, 1969.

Tucker, Rahat W. "Oil: The Issue of American Intervention." and "Further Reflections on Oil and Force." *Commentary,* January 1975, pp. 21-31, and March 1975, pp. 45-46.

Tugenhat, Christopher. *Oil: The Biggest Business.* New York: G.P. Putnam's Sons, 1968.

Tugenhat, Christopher, and Adrian Hamilton. *Oil: The Biggest Business.* 2d ed., rev. London: Ayer Methuen, 1975.

Tugenhat, Christopher. "Political Approach to the World Oil Problem." *Harvard Business Review,* January, 1976, pp. 45-55.

Tugwell, Franklin. *The Politics of Oil in Venezuela.* Palo Alto: Stanford University Press, 1975.

U.S. Government, Federal Energy Administration. *Oil: Possible Levels of Future Production.* Washington, D.C.: Government Printing Office, November 1974.

U.S. Government. *Project Independence.* Washington, D.C.: Federal Energy Administration, November 1974.

U.S. Government. *Report on the International Petroleum Control,* submitted to the Subcommittee on Monopoly of the Select Committee on Small Business, U.S. Senate. Washington, D.C.: U.S. Federal Trade Commission, August, 1952.

U.S. Government. *Multinational Corporations and United States Foreign Policy.* U.S. Senate Committee on Foreign Relations, Subcommittee on Multinational Corporations, *Hearings,* 93rd Congress, second session (Church Committee Hearings). Washington, D.C.: Government Printing Office, 1974.

Vallenilla, Luis. *Oil: The Making of a New Economic Order.* New York: McGraw-Hill, 1975.

Vernon, Raymond, ed. *The Oil Crisis.* New York: W.W. Norton, 1976. (*Daedalus,* Vol. 104, No. 4, Fall 1975.)

Vicher, Ray. *Kingdom of Oil, the Middle East.* New York: Scribner's, 1974.

Watkins, G. Campbell, ed. *Oil in the Seventies.* Vancouver: Fraser Institute, 1977.

Weisberg, Richard C. *The Politics of Crude Oil Pricing in the Middle East 1970-1975.* Berkeley: University of California Press, 1977.

Williams, Maurice. "Aid Programs of OPEC Countries." *Foreign Affairs,* January 1976, pp. 308-324.

Wilrich, Mason. *Energy and World Politics.* New York: Free Press, 1975.

Wilson, Carroll. *Energy Global Prospects, 1985-2000.* New York: McGraw-Hill, 1977.

Windsor, Philip. *Oil: A Guide Through the Total Energy Jungle.* Boston: Gambit, 1976.

Wyant, Frank H. *The United States, OPEC, and Multinational Oil.* Lexington, Mass.: Lexington Books, D.C. Heath, 1977.

Yager, Joseph A., and Eleanor B. Steinberg. *Energy and U.S. Foreign Policy,* (Report of Ford Foundation Energy Policy Project) Cambridge, Mass.: Ballinger, 1974.

Yamani, Ahmed Zaki. *Economics of the Petroleum Industry.* Beirut: Middle East Research and Publishing Center, 1970.

Zartman, I. William, ed. *The 50% Solution.* Garden City: Anchor, 1976.

Index

About the Author

Loring Allen teaches the economics of energy and environmental energy at the University of Missouri-St. Louis. Research on his *Venezuelan Economic Development* (1977) stimulated his interest in OPEC and oil and led to this book. He has written other books—including *Lecciones de Economia Regional, Soviet Economic Warfare,* and *The Formation of American Trade Policy*—as well as many articles. Dr. Allen, who received his doctorate from Harvard University, has also taught at the University of Virginia, at the University of Oregon, and at universities in Spain and Latin America.